GRAVITY SANITARY SEWER DESIGN AND CONSTRUCTION

ASCE MANUALS AND REPORTS ON ENGINEERING PRACTICE NO. 60
WPCF MANUAL OF PRACTICE NO. FD-5

Prepared by a Joint Task Force of the
American Society of Civil Engineers
and the
Water Pollution Control Federation

The Joint Task Force was composed of members of the ASCE Environmental Engineering Division Committee on Water Pollution Management, Pipeline Division Committee on Pipeline Planning, and the WPCF Facilities Development Subcommittee of the Technical Practice Committee

Published by the
American Society of Civil Engineers
345 East 47th Street
New York, New York 10017-2398
and the
Water Pollution Control Federation
601 Wythe Street
Alexandria, VA 22314-1994

The material presented in this publication has been prepared in accordance with generally recognized engineering principles and practices, and is for general information only. This information should not be used without first securing competent advice with respect to its suitability for any general or specific application.

The contents of this publication are not intended to be and should not be construed to be a standard of the American Society of Civil Engineers (ASCE) or the Water Pollution Control Federation (WPCF) and are not intended for use as a reference in purchase specifications, contracts, regulations, statutes, or any other legal document.

No reference made in this publication to any specific method, product, process, or service constitutes or implies an endorsement, recommendation, or warranty thereof by ASCE or WPCF.

ASCE and WPCF make no representation or warranty of any kind, whether express or implied, concerning the accuracy, completeness, suitability or utility of any information, apparatus, product, or process discussed in this publication, and assumes no liability therefor.

Anyone utilizing this information assumes all liability arising from such use, including but not limited to infringement of any patent or patents.

Library of Congress Cataloging in Publication Data
Main entry under title:

Gravity sanitary sewer design and construction.

(ASCE manuals and reports on engineering practice; no. 60) (WPCF manual of practice; no. FD-5)
Includes bibliographical references and index.
1. Sewerage—Design and construction. I. American Society of Civil Engineers. II. Water Pollution Control Federation. III. Series. IV. Series: WPCF manual of practice; no. FD-5.
TD201.W337 no. FD-5 [TD678] 628.1'68 [628'.2] 81-69182
ISBN 0-87262-313-0

Copyright © 1982 by the American Society of Civil Engineers
and the Water Pollution Control Federation,
All Rights Reserved.
Library of Congress Catalog Card No. 81-69182
ISBN 0-87262-313-0
Manufactured in the United States of America.

Joint Task Force on Sanitary Sewers

Dan L. Glasgow, Chairman*
G.H. Abplanalp
T. Asano*
J.B. Butler
D.G. Chase*
S.R. Copas
L.W. Curtis
D.M. Fuller
S.A. Gill
W.E. Goodnow
F.A. Grant
M.J. Hecker
J.C. Hill
J.P. Hood
S. Koplowitz*
P.A. Kuhn*
J.R. Loveland*
H.H. Lund

C.T. Mallder
A.J. Messina
R.T. McLaughlin*
T.E. McMahon
J.D. Morrow
T.E. Mosure*
W.D. Nesbeitt*
H.T. Peck
B.L. Pickard
R.D. Pomeroy*
H.F. Reheis
J.G. Richard, Jr.
B.J. Schrock
J.F. Seidensticker
P.M. Shepard, Jr.
C.F. Steiner*
L.E. Stockton

*Principal contributing author

In addition to the Task Force, other reviewers included:

R.J. Barletta
W.C. Boyle
M.A. Collins
K.K. Kienow
J.G. Hendrickson
L.P. Jaworski

O. Milgram
S. Mruk
H. Peistrup
J. Roberts
H.E. Schmidt
R.P. Walker, Jr.

The Manual was also reviewed by the WPCF Wastewater Collection System Committee that included:

J.G. Johanson, Chairman
R. Berry
M. Clark
D. Horsefield

R. Hyams
R. McMahon
J. Zaffle

Staff assistance was provided by Eugene D. De Michele for the Water Pollution Control Federation and Harry N. Tuvel for the American Society of Civil Engineers.

MANUALS AND REPORTS ON ENGINEERING PRACTICE

(As developed by the ASCE Technical Procedures Committee, July, 1930, and revised March, 1935, February, 1962, April, 1982)

A manual or report in this series consists of an orderly presentation of facts on a particular subject, supplemented by an analysis of limitations and applications of these facts. It contains information useful to the average engineer in his everyday work, rather than the findings that may be useful only occasionally or rarely. It is not in any sense a "standard," however; nor is it so elementary or so conclusive as to provide a "rule of thumb" for nonengineers.

Furthermore, material in this series, in distinction from a paper (which expresses only one person's observations or opinions), is the work of a committee or group selected to assemble and express information on a specific topic. As often as practicable the committee is under the general direction of one or more of the Technical Divisions and Councils, and the product evolved has been subjected to review by the Executive Committee of that Division or Council. As a step in the process of this review, proposed manuscripts are often brought before the members of the Technical Divisions and Councils for comment, which may serve as the basis for improvement. When published, each work shows the names of the committee by which it was compiled and indicates clearly the several processes through which it has passed in review, in order that its merit may be definitely understood.

In February, 1962, (and revised in April, 1982) the Board of Direction voted to establish:

> A series entitled 'Manuals and Reports on Engineering Practice,' to include the Manuals published and authorized to date, future Manuals of Professional Practice, and Reports on Engineering Practice. All such Manual or Report material of the Society would have been refereed in a manner approved by the Board Committee on Publications and would be bound, with applicable discussion, in books similar to past Manuals. Numbering would be consecutive and would be a continuation of present Manual numbers. In some cases of reports of joint committees, bypassing of Journal publications may be authorized.

ABSTRACT

GRAVITY SANITARY SEWER DESIGN AND CONSTRUCTION

This Manual provides both theoretical and practical guidelines for the design and construction of gravity sanitary sewers.

The initial chapter introduces the organization and administrative phases of the sanitary sewer project. Subsequent chapters are presented in a sequence detailing the parameters necessary to establish the design criteria, complete the design and award a construction contract. The Manual concludes with a discussion of the commonly used methods of sanitary sewer construction.

The Manual is intended to be of practical use to the designer of a gravity sanitary sewer system and is based upon the experience of engineers in the field of sanitary sewer structural and hydraulic design. Charts, illustrations and example problem solutions are used liberally throughout to reinforce the text.

ASCE — Manual and Reports on Engineering Practices No. 60
WPCF — Manual of Practice — No. FD-5
ERRATA

1. **Page 54,** Equation 4.5 should read:

$$\frac{d[s]}{dt} = \frac{3.28M\,[EBOD]}{0.25D}(1 + 0.12D)$$

2. **Page 55,** 6th line, change feet to metres.

3. **Page 55,** add Equation 4.7a:

$$[S]_{lim} = \frac{0.52 \times 10^{-3}\,[EBOD]}{(sv)^{0.375}} \times \frac{dm}{r}$$

4. **Page 55,** add Equation 4.8a:

$$[S]_2 = [S]_{lim} - \left\{ \frac{[S]_{lim} - [S]_1}{\log^{-1}\left[\dfrac{(sv)^{0.375}\,\Delta t}{2.30\,dm}\right]} \right\}$$

5. **Page 61,** 3rd line, change 0.12 to 0.012

6. **Page 61,** 6th line, change B to b

7. **Page 84,** Equation 5.15 should read:

$$\frac{h_f}{l} = \frac{fv^2}{2gD}$$

8. **Page 107,** Table 5-2, change: 0.50;   to: 0.49;
                                   0.10;   0.34;
                                   0.28;   0.25;
                                   0.22; and   0.20; and
                                   0.15   0.13

9. **Page 169,** 2nd line, change kilograms to newtons

10. **Page 180,** 2nd line after Equation 9.10, change kilograms to newtons

11. **Page 184,** bottom line, change kilograms to newtons

12. **Page 220,** 4th line after Equation 9.22, word should be "equations".

Contents

Chapter 1	**Organization and Administration of Sanitary Sewer Projects**	
A.	Introduction	1
B.	Definition of Terms and Classification of Sanitary Sewers	1
C.	Phases of Project Development	2
	1. Preliminary or Investigative	2
	2. Design	3
	3. Construction	3
	4. Operation	3
D.	Interrelations of Project Development Phases	3
E.	Parties Involved in Design and Construction of Sanitary Sewer Projects	4
	1. Owner	4
	2. Engineer	5
	3. Contractor	5
	4. Other Parties	5
F.	Role of Parties in Each Phase	6
	1. Preliminary or Investigative	6
	2. Design	6
	3. Construction	6
	4. Operation	6
G.	Control of Sanitary Sewer System Use	7
H.	Federal and State Planning and Funding Assistance	7
	1. Federal Assistance	7
	2. State Assistance	8
I.	Local Funding	8
	1. General Obligation Bonds	8
	2. Special Assessment Bonds	8
	3. Revenue Bonds	8
	4. Pay-As-You-Go Financing	9
	5. Revenue Programs and Rate Setting	9
	a. Ad Valorem Taxes	9
	b. Service charges	9
	c. Connection Charges	10
	d. Combination Taxes and Service Charges	10
J.	Safety	10
	1. Investigations	10
	2. Design	10
	3. Construction	11
	4. Operation and Maintenance	11
K.	National Environmental Policy Act of 1969	11
L.	Measurement Units	12
M.	References	14
Chapter 2	**Surveys and Investigations**	
A.	Introduction	15
B.	Types of Information Required	15
	1. Physical	15
	2. Developmental	16
	3. Political	16
	4. Sanitary	16
	5. Financial	16

	C.	Sources of Information	17
		1. Physical	17
		2. Developmental	17
		3. Political	17
		4. Sanitary	17
		5. Financial	17
	D.	Surveys for Different Project Phases	18
		1. Preliminary Surveys	18
		2. Design Surveys	18
		3. Construction Surveys	19
	E.	Investigations	19
Chapter 3		**Quantity of Wastewater**	
	A.	Introduction	21
	B.	Design Period	21
	C.	Design Flows	22
		1. Population Estimates	23
		2. Land Use	24
		3. Tributary Areas	24
		4. Per Capita Wastewater Flow	25
		5. Flow Estimates Based on Population and Flow Trends	30
		6. Distribution of Population	30
		7. Commercial Contributions	31
		8. Industrial Contributions	33
		9. Institutional Contributions	34
		10. Cooling Water	34
		11. Infiltration/Inflow Contributions	35
		a. Infiltration from Leakage into Manholes; Roof and Areaway Drainage; Surface Runoff into Basements and Crawl Spaces	35
		b. Infiltration	36
		12. Minimum and Peak Flows of Sanitary Wastewater	38
		a. Fixture-Unit Method of Design	40
	D.	Recapitulation	44
	E.	References	46
Chapter 4		**Sulfide Generation, Corrosion and Corrosion Protection in Sanitary Sewers**	
	A.	Introduction	47
	B.	Sulfide Generation	47
	C.	Forecasting Sulfide Conditions	49
		1. Major Design Considerations—Sulfide Generation	49
		2. Sulfide Forecast Criteria	51
		a. Flow-Slope Relationship	51
		b. Potential for Sulfide Buildup	53
		c. Quantitative Sulfide Forecasting—Filled Pipe Condition	54
		d. Quantitative Sulfide Forecasting—Partially Filled Pipe Conditions	54
	D.	In-Line Sulfide Studies	55
	E.	Design Considerations—Effects of Sulfides	56
	F.	Sulfide Control	57
	G.	Other Corrosive Conditions	58

H.		Corrosion Protection	59
	1.	Linings and Coatings	59
	2.	Composition of Materials and/or Thickness of Pipe Material	60
I.		References	66

Chapter 5   Hydraulics of Sewers

A.			Introduction	67
B.			Terminology and Symbols	68
C.			Hydraulic Principles	70
	1.		Types of Flow	70
	2.		Examples of Flow Situations Encountered	70
	3.		Equations of Motion	78
		a.	Analysis of One-Dimensional Steady Flow	78
		b.	Continuity Principle	78
		c.	Energy Principle	80
		d.	Momentum Principle	82
	4.		Methods of Application and Calculation	82
		a.	Friction Losses	82
		b.	Other Energy Losses	85
		c.	Specific Energy and Alternate Depths	86
		d.	Momentum Equation	88
		e.	General Features of Water-Surface Profiles	91
D.			Flow Resistance	93
	1.		Friction Formulas	93
		a.	General	93
		b.	Kutter and Manning Formulas	94
		c.	Hazen-Williams Formula	97
		d.	Darcy-Weisbach Equations	98
	2.		Factors Affecting Friction Coefficients	100
		a.	General	100
		b.	Conduit Material	101
		c.	Size of Conduit	101
		d.	Conduit Shape	101
		e.	Depth of Flow	101
E.			Design Computation	103
	1.		Application of Hydraulic Computations	103
	2.		Capacity and Flow Estimates	103
		a.	Full Pipes	103
		b.	Partly Full Pipes	103
		c.	Self-Cleansing Velocities	105
	3.		Water-Surface Elevations at Important Locations	107
		a.	Critical Depth	107
		b.	Hydraulic Jump	108
		c.	Appurtenances and Controls	108
	4.		Detailed Analyses	110
		a.	Water-Surface Profiles	110
		b.	Flow Routing and Wave Calculations	110
F.			References	110

Chapter 6   Design of Sanitary Sewer Systems

A.	Introduction	112
B.	Energy Concepts of Sewer Systems	112

C.	Combined vs. Separate Sewers	113
D.	Layout of System	113
E.	Curved Sanitary Sewers	115
	1. Rigid Pipe	115
	2. Flexible Pipe	119
F.	Type of Conduit	121
G.	Ventilation	121
H.	Depth of Sanitary Sewer	121
I.	Flow Velocities and Design Depths of Flow	122
J.	Infiltration/Inflow	123
K.	Infiltration/Exfiltration and Low-Pressure Air Testing	123
	1. Infiltration-Exfiltration Test Allowance	123
	2. Infiltration-Exfiltration Testing	123
	3. Low-Pressure Air Testing	124
L.	Design for Various Conditions	124
	1. Open Cut	124
	2. Tunnel	125
	3. Sanitary Sewers Built in Rock	125
	4. Exposed Sanitary Sewers	125
	5. Foundations	125
	6. Sanitary Sewers on Steep Slopes	125
M.	Relief Sewers	128
N.	Organization of Computations	128
O.	References	129

Chapter 7   **Appurtenances and Special Structures**

A.	Introduction	130
B.	Manholes	130
	1. Objectives	130
	2. Manhole Spacing and Location	130
	3. General Shape and Dimensions	130
	4. Shallow Manholes	132
	5. Construction Material	132
	6. Frame and Cover	132
	7. Connection Between Manhole and Sewer	133
	8. Steps	133
	9. Channel and Bench	134
	10. Manholes on Large Sewers	135
C.	Bends	136
D.	Junctions	137
E.	Drop Manholes	137
F.	Terminal Cleanouts	138
G.	Service Laterals	139
H.	Check Valves and Relief Overflows	141
I.	Siphons	143
	1. Single- and Multiple-Barrel Siphons	143
	2. Profile	144
	3. Air Jumpers	144
	4. Sulfide Generation	145
J.	Flap or Backwater Gates	145
K.	Sewers Above Ground	145
L.	Underwater Sewers and Outfalls	146
	1. Ocean Outfalls	146
	2. Other Outlets	147

M.	Measuring Wastewater Flows 148
N.	References 152

Chapter 8 Materials for Sewer Construction

- A. Introduction 153
- B. Sewer Pipe Materials 154
    1. Rigid Pipe 154
        - a. Asbestos Cement Pipe (ACP) 154
        - b. Cast Iron Pipe (CIP) 154
        - c. Concrete Pipe 155
        - d. Vitrified Clay Pipe (VCP) 156
    2. Flexible Pipe 157
        - a. Ductile Iron Pipe (DIP) 157
        - b. Steel Pipe 157
        - c. Thermoplastic Pipe 158
        - d. Thermoset Plastic Pipe 161
- C. Pipe Joints 162
    1. General Information 162
    2. Types of Pipe Joints 162
        - a. Gasket Pipe Joints 163
        - b. Bituminous Pipe Joints 163
        - c. Cement Mortar Pipe Joint 163
        - d. Elastomeric Sealing Compound Pipe Joints 163
        - e. Solvent Cement Pipe Joints 163
        - f. Heat Fusion Pipe Joints 164
        - g. Mastic Pipe Joints 164
        - h. Sealing Band Joints 164
- D. Summary 164
- E. References 164

Chapter 9 Structural Requirements

- A. Introduction 166
- B. Loads on Sewers Caused by Gravity Earth Forces 166
    1. General Method—Marston Theory 166
    2. Types of Loading Conditions 167
    3. Loads for Trench Conditions 168
        - a. Use of Marston's Formula 168
        - b. Soil Characteristics—Trench Conditions 174
    4. Loads for Embankment Conditions 175
        - a. Positive Projecting Sewer Pipe 175
        - b. Negative Projecting and Induced Trench Sewer Pipes 179
        - c. Sewer Pipe Under Sloping Embankment Surfaces 182
    5. Loads for Jacked Sewer Pipe and Certain Tunnel Conditions 183
        - a. Load-Producing Forces 183
        - b. Marston's Formula 184
        - c. Tunnel Soil Characteristics 185
        - d. Effect of Excessive Excavation 186
    6. Loads for Tunnels 186
        - a. Load-Producing Forces 187
    7. Alternate Design Method 187

C.		Superimposed Loads on Sanitary Sewers 188	
	1.	General Method 188	
	2.	Concentrated Loads 189	
	3.	Impact Factor 189	
	4.	Distributed Loads 191	
	5.	Sewer Pipe Under Railway Tracks 192	
	6.	Sewer Pipe Under Rigid Pavement 192	
D.		Pipe Bedding and Backfilling 193	
	1.	General Concepts 193	
	2.	Foundation 194	
	3.	Bedding 194	
	4.	Haunching 195	
	5.	Initial Backfill 196	
	6.	Final Backfill 196	
E.		Design Safety Factor and Performance Limits 196	
	1.	General Concepts 196	
	2.	Rigid Sewer Pipe 197	
	3.	Flexible Sewer Pipe 197	
	4.	Recommendations for Field Procedures 198	
		a. Effect of Trench Sheeting 198	
		b. Trench Boxes 199	
		c. Pipe Bedding and Embedment 199	
F.		Rigid Sewer Pipe Design 200	
	1.	General Relationships 200	
	2.	Laboratory Strength 200	
	3.	Design Relationships 201	
	4.	Rigid Sewer Pipe Installation—Classes of Bedding and Bedding Factors for Trench Conditions 202	
		a. Class A—Concrete Cradle 202	
		Class A—Concrete Arch 203	
		b. Class B Bedding 204	
		c. Class C Bedding 204	
		d. Class D Bedding 204	
	5.	Encased Pipe 204	
	6.	Field Strength in Embankments 206	
		a. Positive Projecting Sewer Pipe 206	
		b. Negative Projecting Sewer Pipe 209	
		c. Induced Trench Conditions 209	
G.		Flexible Sewer Pipe Design 209	
	1.	General Method 209	
	2.	Design of Plastic Sewer Pipes 212	
		a. Laboratory Load Test 213	
		b. Design Relationships 214	
		c. Loads on Flexible Plastic Pipe 215	
		d. Field Deflection of Flexible Plastic Pipe 215	
	3.	Plastic Sewer Pipe Installation 216	
		a. Bedding, Haunching and Initial Backfill 216	
		b. Final Backfill 217	
	4.	Soil Classification 217	
		a. Class I 217	
		b. Class II 217	
		c. Class III 218	
		d. Class IV and V 218	
	5.	Design of Ductile Iron Sewer Pipes 218	

		6.	Design of Corrugated Metal Sewer Pipes 220
		7.	Flexible Sewer Pipe Installation 221
			a. Bedding, Haunching and Initial Backfill 221
			b. Final Backfill 221
	H.		References 221
Chapter 10		**Construction Contract Documents**	
	A.		Introduction 223
	B.		Contract Drawings 224
		1.	Purpose 224
		2.	Field Data 224
		3.	Preparation 225
		4.	Contents 226
			a. Arrangement 226
			b. Title Sheet 226
			c. Title Blocks 226
			d. Index/Legend 227
			e. Location Map 227
			f. General Notes 227
			g. Subsoil Information 227
			h. Survey Control and Data 227
			i. Sewer Plans 229
			j. Sewer Profile 229
			k. Sewer Sections 229
			l. Sewer Details 229
			m. Special Details 233
		5.	Record Drawings 234
	C.		Project Manual 234
		1.	Introduction 234
		2.	Purpose 234
		3.	Arrangement 235
		4.	Addenda 235
		5.	Bidding Requirements 236
			a. Invitation to Bid 236
			b. Instructions to Bidders 236
			c. Bid Form 236
		6.	Contract Forms 237
			a. Form of Contract 237
			b. Surety Bonds 237
			c. Special Forms 237
		7.	Conditions of Contract 237
		8.	Specifications 238
			a. General Requirements (CSI Division 1) 238
			b. Material and Workmanship Specifications (CSI Divisions 2 to 16) 238
		9.	Modifications 239
		10.	Supplementary Information 239
		11.	Standard Specifications 239
		12.	Project Manual Check List 240
	D.		References 243
Chapter 11		**Construction Methods**	
	A.		Introduction 244

- B. Construction Surveys 244
  1. General 244
  2. Preliminary Layouts 244
  3. Setting Line and Grade 245
  4. Tunnel Construction 246
- C. Site Preparation 247
- D. Open-Trench Construction 247
  1. Trench Dimensions 247
  2. Excavation 248
     a. Stripping 248
     b. Drilling and Blasting 248
     c. Trenching 249
  3. Sheeting and Bracing 251
  4. Dewatering 253
  5. Foundations 253
  6. Pipe Sanitary Sewers 254
     a. Sewer Pipe Quality 254
     b. Sewer Pipe Handling 254
     c. Sewer Pipe Placement 254
  7. Backfilling 255
     a. General Considerations 255
     b. Degree of Compaction 255
     c. Methods of Compaction 256
     d. Backfilling Sequence 257
  8. Surface Restoration 258
- E. Tunneling 258
  1. General Classification 258
  2. Auger or Boring 259
  3. Jacking 259
     a. Alignment 259
     b. Continuous Sewers by Jacking 260
  4. Mining Methods 260
     a. Tunnel Shields 260
     b. Boring Machines 261
     c. Open-Face Mining Without Shields 261
     d. Primary or Temporary Lining 261
     e. Oval Precast Concrete Rings 264
     f. Tunnel Excavating Equipment 264
     g. Shafts 264
     h. Main Shafts and Emergency Exits 264
     i. Compressed Air Equipment and Locks 266
     j. Ventilating Air 266
- F. Special Construction 266
  1. Railroad Crossings 266
  2. Crossing of Principal Traffic Arteries 267
  3. Stream and River Crossings 267
     a. Sanitary Sewer Crossing Under Waterway 267
     b. Sanitary Sewer Crossing Spanning Waterway 268
  4. Outfall Structures 268
     a. Riverbank Structures 268
     b. Ocean Outfalls 269
- G. Sewer Appurtenances 270
- H. Construction Records 270
- I. References 270

Index 273

FOREWORD

In 1960, a joint committee of the Water Pollution Control Federation and the American Society of Civil Engineers published the Manual of Practice on the *Design and Construction of Sanitary and Storm Sewers*. In 1964, a second joint committee was formed to revise and expand the Manual; in 1969 the revised edition was published. In subsequent reprintings, the 1969 edition of the Manual was continuously revised to provide information on improved and more current practices.

In 1978 the Water Pollution Control Federation authorized preparation of this Manual of Practice devoted to Gravity Sanitary Sewers. In 1979 the American Society of Civil Engineers entered into an agreement with the Water Pollution Control Federation to continue their joint publication relationship.

This Manual should be considered by the practicing engineer as an aid and a check list of items to be considered in a gravity sanitary sewer project, as represented by acceptable current procedure. It is not intended to be a substitute for engineering experience and judgment, or as a treatise replacing standard texts and reference material.

In common with other manuals prepared on special phases of engineering, the Manual recognizes that this field of engineering is constantly progressing with new ideas, materials and methods coming into use. Other alternatives available to the designer of sanitary sewers include vacuum, pressure, vacuum-pressure and small diameter gravity sewers. It is hoped that users will present any suggestions for improvement to the Technical Practices Committee of the Water Pollution Control Federation or the Environmental Engineering and Pipeline Divisions of the American Society of Civil Engineers for possible inclusion in future revisions to keep this Manual current.

The members of the Committee thank the reviewers of the Manual for their assistance in submitting their suggestions for improvement.

CHAPTER 1

ORGANIZATION AND ADMINISTRATION OF SANITARY SEWER PROJECTS

A. INTRODUCTION

A major capital investment made by a community is its sanitary sewer system. The system's function is vaguely recognized and sewers are seldom seen by the public — except for the manhole covers. Sanitary systems are essential to protecting the public health and welfare in all areas of concentrated population and development. Every community produces wastewater of domestic, commercial, and industrial origin. Sanitary sewers perform the vitally needed functions of collecting these wastewaters and conveying them to points of treatment and disposal.

The various stages of design and construction of sanitary sewer projects require an understanding of the objectives of each stage of the project and of the responsibilities and interests of the parties involved.

Separate sanitary and storm sewers are highly desirable and used with few exceptions in new systems. The major advantages of separate systems, including wastewater treatment plants, are the protection of watercourses from pollution and the exclusion of stormwater from the treatment system with a consequent saving in treatment plant construction and operating cost. Combined sewers frequently are encountered in older communities where it may be extremely difficult or costly to provide separate systems. Separation is desired, where economically feasible, to reduce the magnitude of facilities and energy demand of treatment works.

B. DEFINITION OF TERMS AND CLASSIFICATION OF SANITARY SEWERS

The following terms as used in this Manual are defined in the *Glossary-Water and Wastewater Control Engineering* (1) as follows:

waste water — In a legal sense, water that is not needed or which has been used and is permitted to escape, or which unavoidably escapes from ditches, canals, or other conduits, or reservoirs of the lawful owners of such structures. See also **wastewater.**

wastewater — The spent or used water of a community or industry which contains dissolved and suspended matter.

sanitary sewer — A sewer that carries liquid and waterborne wastes from residences, commercial buildings, industrial plants, and institutions, together with minor quantities of ground, storm, and surface waters that are not admitted intentionally. See also **wastewater.**

combined sewer — A sewer intended to receive both wastewater and storm or surface water.

relief sewer — (1) A sewer built to carry flows in excess of the capacity of an existing sewer. (2) A sewer intended to carry a portion of the flow from a district in which the existing sewers are of insufficient capacity, and thus prevent overtaxing the latter.

building sewer — In plumbing, the extension from the building drain to the public sewer or other place of disposal. Also called house connection.

building drain — In plumbing, that part of the lowest horizontal piping within a building that conducts water, wastewater, or storm water to a building sewer.

lateral sewer — a sewer that discharges into a branch or other sewer and has no other common sewer tributary to it.

main sewer — (1) In larger systems, the principal sewer to which branch sewers and submains are tributary; also called trunk sewer. In small systems, a sewer to which one or more branch sewers are tributary. (2) In plumbing, the public sewer to which the house or building sewer is connected.

trunk sewer — A sewer that receives many tributary branches and serves a large territory.

intercepting sewer — A sewer that receives dry-weather flow from a number of transverse sewers or outlets and frequently additional predetermined quantities of storm water (if from a combined system), and conducts such waters to a point for treatment or disposal.

outfall — (1) The point, location, or structure where wastewater or drainage discharges from a sewer, drain, or other conduit. (2) The conduit leading to the ultimate disposal area.

outfall sewer — A sewer that receives wastewater from a collecting system or from a treatment plant and carries it to a point of final discharge.

separate sewer — A sewer intended to receive only wastewater or storm water or surface water. See also **combined sewer, sanitary sewer, storm sewer.**

separate sewer system — A sewer system carrying sanitary wastewater and other waterborne wastes from residences, commercial buildings, industrial plants, and institutions, together with minor quantities of ground, storm, and surface waters that are not intentionally admitted. See also **wastewater, combined sewer.**

storm sewer — A sewer that carries storm water and surface water, street wash and other wash waters, or drainage, but excludes domestic wastewater and industrial wastes. Also called storm drain.

C. PHASES OF PROJECT DEVELOPMENT

Conception and development of typical sanitary sewer projects comprise the following phases:

1. Preliminary or Investigative

The objective of this phase is to establish the broad technical and economic bases for environmental assessments, policy decisions and final designs. The importance of this phase cannot be overemphasized. Inadequate preliminary work will be detrimental to all succeeding phases and may endanger the successful completion of the project or cause the owner to undertake planning which may not produce the most economical or efficient result. This phase usually culminates in an engineering report which includes items such as:

(a) Statement of the problem and review of existing conditions.

(b) Capacities and conditions required to provide service for design period.
(c) Method of achieving the required service — if more than one method is available, an evaluation of each alternative method.
(d) General layouts of the proposed system with indication of stages of development to meet the ultimate condition when the project warrants stage development.
(e) Establishment of applicable engineering criteria and preliminary sizing and design that will permit preparation of construction and operating cost estimates of sufficient accuracy to provide a firm basis for feasibility determination, financial planning, and consideration of alternative methods of solution.
(f) Various available methods of financing and their applicabiity to the project.
(g) An assessment of the anticipated environmental impacts of construction and review of the long-term effects on the environment. The environmental assessment when required must be sufficiently thorough and objective to enumerate the environmental consequences of the project. Measures to mitigate any negative impacts should be set forth. Projects which have community-wide consequences and which are supported by federal funds may be subjected to detailed evaluation through Federal Environmental Impact Statement procedures.

It must be recognized that the preliminary engineering report is not a detailed working design or plan from which a sanitary sewer project can be constructed. Indeed, such detail is not necessary to meet the objectives of the preliminary or investigative phase or the environmental assessment. Nonetheless, proper preliminary engineering is the fundamental initial step of final planning. (For additional information concerning the Survey and Investigation phases of sanitary sewer project development, See Chapter 2.)

2. Design

The design phase of a sanitary sewer project comprises the preparation of construction plans and specifications. These documents form the basis for bidding and performance of the work; they must be clear and concise. Design, therefore, consists of the elaboration of the preliminary plan to include all details necessary to construct the project.

3. Construction

This phase involves the actual building of the project according to the plans and specifications previously prepared.

4. Operation

Although this manual is devoted to matters of design and construction, the operation of a sanitary sewer system is an important aspect in the development of such projects.

D. INTERRELATIONS OF PROJECT DEVELOPMENT PHASES

Since all phases of sanitary sewer projects are interrelated, the following points are applicable:
(1) The capacity, arrangement, and details of a sanitary sewer system will not be satisfactory unless the preliminary or investigative phase is current

and properly completed.

(2) Adequate preliminary engineering and estimating are essential to sound financial planning, without which subsequent phases of the project may be placed in jeopardy.

(3) Environmental assessment documentation is intended to provide a single source of comparison and evaluation of all development, construction, and operation phases of the project.

(4) Inadequate design or improperly prepared plans and specifications can lead to confusion in construction, higher costs, failure of the project to meet intended functions, or actual structural or hydraulic failure of component parts.

(5) Proper execution of the construction phase is necessary to produce the quality and features provided by adequate design. Moreover, the value of the design can be lost by incompetent or careless handling of the construction phase.

(6) All sanitary sewer projects have certain features requiring operation and maintenance. Unless they are anticipated and provided for, the usefulness of the project will be impaired.

E. PARTIES INVOLVED IN DESIGN AND CONSTRUCTION OF SANITARY SEWER PROJECTS

Engineering projects, including sanitary sewers, are the result of the combined efforts of the several interested parties. The owner, engineer, and contractor are the principal participants. The project attorney, financial consultant, various regulatory agencies, and other specialists also are involved to varying degrees. Responsibilities of these individuals or organizations are summarized as follows:

1. Owner

The owner's needs initiate the project and he provides the necessary funds. The owner is party to all contracts for services and construction and may act directly or through any duly authorized agent. The owner most often is the collective citizens of a governmental unit whose affairs may be handled by various legislative and administrative bodies. The owner may also be a private group.

When the owner is a governmental unit, its business may be conducted by one of the following, depending on the organization of the unit and the laws controllings its operations:

(a) City councils or similar bodies, carrying out sanitary sewer projects as only one of many duties for the given unit.

(b) A special commission or board of a governmental unit dealing with more limited areas of interest than usually are handled by a city council. Such boards or commissions may be concerned with sanitary sewer projects alone or with a governmental unit's general utility system. The geographical limits of responsibility of such boards or commissions coincide with those of the parent governmental unit.

(c) A specially established district, agency, or authority whose geographical limits are unique and whose affairs are administered by a separate and distinct administrative board or commission. Such units commonly are referred to as "districts"—for example, the County Sanitation Districts of Los Angeles County, which includes 74 separate cities and the unincorporated county. Often, the responsibilities of such sewering districts are limited to main trunk sanitary sewers, intercepting sanitary sewers, treatment facilities, and outfalls, leaving lateral sewers as the responsibility of the individual governmental units within the area served by the larger district.

The fund-raising powers of the first two bodies are usually regulated by the same laws which apply to financing by the parent governmental unit. Fund-raising powers for a specially constituted district may show considerable variation due to the many differences in legislative provisions for special district formation and financing.

Temporary private ownership of sanitary sewer projects sometimes is encountered in new developments. The developer may construct the sanitary sewer system and later transfer title to the appropriate governmental unit in accordance with local regulation. In some instances, wastewater systems and treatment works are under the permanent ownership of private utility companies.

2. Engineer

The engineer has the responsibility of supplying the owner with the basic information needed to make project implementation policy decisions, detailed plans and specifications necessary to bid and construct the project, consultation, general and residential inspection during construction, and services necessary for the owner to establish satisfactory operation and maintenance procedures. The engineer's responsibilities are all of a professional character and must be discharged in accordance with ethical standards by qualified engineering personnel.

Engineering for sanitary sewer projects may be performed either by engineering departments which are a part of a governmental unit or by private engineering firms retained by the owner for specific projects. In many instances, sanitary sewer projects are a joint effort by both types of organizations.

3. Contractor

The contractor performs the actual construction work under the terms of the contract documents prepared by the engineer. The construction agreement is between the contractor and the owner. One or more contractors may perform the work on a single project.

The functions of the contractor may be carried out by an owner's employees especially organized for construction purposes, but such practice for sanitary sewer projects of any magnitude is not common.

4. Other Parties

Others who may enter into sewer project development, particularly in the United States, are as follows:

(a) Legal Counsel—All public works projects are subject to local and state laws; competent legal advice is required to assure compliance with these laws and the avoidance of setbacks because of legal defects in the project. Special legal counsel may also be required in connection with financing the project, particularly where a bond issue is involved.

(b) Financial Consultant—Advisory services with respect to project financing often are required and may be provided as a separate and specialized service. Such services occasionally are provided as part of a general financing agreement with a financing agency.

(c) Regulatory Agencies—The most frequently encountered regulatory body is the state health department, department of environmental quality or in some states a specially designated water pollution control agency which usually adopts minimum standards pertaining to features of design, plans, and specifications for sanitary sewer projects. Other regulatory bodies having jurisdiction may include agencies such as municipality or sanitary sewer districts; local, regional, or state planning commissions; federal agencies concerned with water pollution control; and federal or state agencies having functional control of navigable waters.

(d) Pipe Manufacturers—These manufacturers produce the specified pipe materials and often will embark on focused developmental research to prove or improve a specific pipe product.

F. ROLE OF PARTIES IN EACH PHASE

The roles of the owner, engineer, and contractor with respect to each other in the different phases of the project are distinct. Unauthorized assumption of roles and duties of one of the parties may result in delays, failures, and/or contractual controversies.

1. Preliminary or Investigative

The owner and the engineer are the principal parties involved in the preliminary phase of sanitary sewer projects. It must be recognized that all policy decisions relating to the project, arranging for financing, etc., rest solely in the hands of the owner.

2. Design

The design phase, up to the time of soliciting and receiving construction bids, involves both the owner and the engineer. Designs prepared by the engineer are normally subject to the approval of the owner. The engineer may recognize preferences of the owner and be guided by these preferences when they are consistent with good engineering practice. The engineer must recognize and conform to the legal, procedural, and regulatory requirements governing each project.

3. Construction

In his relationship with the contractor, the engineer must exercise authority on behalf of the owner. The engineer determines if the work is substantially in accordance with the requirements of the contract documents. He must avoid direct supervision of the contractor's construction operations. If he does control or direct the acts of the contractor, he may become involved as a third party in any legal action brought against the contractor.

4. Operation

Full information on the intended functioning of all parts of a project should be furnished to the owner by the engineer. The owner's staff must assume final responsibility for operation at the time the project or any part of it is completed and accepted by him. In some cases, the engineer, by special agreement with the owner, may provide advisory services in connection with operation and maintenance procedures for a period of time after initial operation.

G. CONTROL OF SANITARY SEWER SYSTEM USE

Of all public utilities, sanitary sewer systems are probably the most abused through misuse. This situation results from a misconception that a sanitary sewer can be used to carry away any unwanted substance or object that can be put into it. The absence of adequate regulations setting forth proper uses and limitations of the system and the lack of enforcement of existing regulations by those responsible for operation of the system, tend to foster such a misconception. Abuse of the sanitary sewer system can result in extensive damage and compound the problems of wastewater treatment. Without proper maintenance and control, a sanitary sewer system may become a hazard to the public safety and increase operating costs unnecessarily.

The following are common consequences of sanitary sewer system misuse:

(1) Explosion and fire hazards resulting from discharge of explosive or flammable substances into the sanitary sewer.

(2) Sanitary sewer clogging by accumulations of grease, bed load, and miscellaneous debris.

(3) Physical damage to sanitary sewer systems resulting from discharge of corrosive or abrasive wastes.

(4) Surface and groundwater overload resulting from improper connections to sanitary sewers.

(5) Watercourse pollution resulting from discharge of wastewater to storm sewers.

(6) Interference with wastewater treatment resulting from extreme wet-weather flows or from wastes not amenable to normal treatment processes.

In the organization of sanitary sewer projects, as well as the management of completed systems, provision must be made for the controlled use of the sanitary sewers and enforcement of appropriate regulations. A comprehensive report on sanitary sewer ordinances is contained in WPCF Manual of Practice No. 3(2). This publication will be helpful in checking the adequacy of existing regulations or in preparing new ones.

H. FEDERAL AND STATE PLANNING AND FUNDING ASSISTANCE

1. Federal Assistance

Federal programs to provide financial assistance for the construction of publicly owned wastewater treatment and transport systems began in the middle 1950s and have grown to major significance. In many states, federal financial assistance programs for qualified wastewater systems are supplemented by state grant and loan programs. The engineer is frequently responsible for locating such sources of financial assistance and preparing the necessary applications for the owner.

The Water Pollution Control Act of 1972, amended as the Clean Water Act of 1977 and 1981, is the most significant source of federal assistance. This law initially required state water pollution control agencies and certain substate agencies to conduct planning for water quality management. It now provides federal grants for the construction costs of wastewater treatment facilities. The grant program is administered by the United States Environmental Protection Agency (EPA) and by certain state water pollution control agencies which have been delegated the authority to administer the program. Availability of financial assistance for sanitary sewers varies from time to time and from state to state, since it is subject to the priority systems of the federal government and of each state for the distribution of federal funding allotments.

The second most significant source of federal assistance is the Farmers Home Administration (FmHA). Grants and loans for such systems are available primarily to small, rural communities. Other federal grants and loans are available from the Department of Housing and Urban Development (HUD), the Economic Development Administration (EDA), and a variety of regional agencies such as the Appalachian Regional Commission and the Coastal Plains Regional Commission.

Federal assistance programs have provided significant aid to local wastewater agencies, but they are often limited by federal appropriations and generally have substantial requirements and administrative procedures which must be followed. A review of current funding allocations and regulations is necessary to determine availability of federal funds at any given time.

2. State Assistance

A majority of the fifty states operate programs to assist local governments in the planning and financing of wastewater projects including sanitary sewers. Many states have area planning and development commissions which prepare or assist in the preparation of planning documents for sanitary sewers for communities within their jurisdiction. States operate grant and loan programs ranging from the very modest up to 100% of the cost of wastewater projects when supplemented by various federal assistance programs. Such programs are usually administered by the state water pollution control agencies or health departments. The current status of state and local programs must be reviewed as a part of project design activities.

I. LOCAL FUNDING

Local funding methods include the following:

1. General Obligation Bonds

These bonds are backed by the full faith and credit of the issuer and often are paid for by the levy of general property taxes. In addition to or instead of taxes, revenues from service charges or other sources may be used to meet bond payments. Advantages of these bonds over other bonds include lower interest rates because of the substantial security provided, and the ease with which they may be sold. However, in most states the issuance of general obligation bonds by local government is subject to constitutional and statutory limitations and to voter approval.

2. Special Assessment Bonds

Such bonds are payable from the receipts of special benefits assessments against certain properties or recipients of benefit. Assessments that are not paid become a lien on the property. Assessments may be based on front footage, area of parcels of property, or other bases. In most cases, the prior consent of landowner's representing some statutory percentage of the total property to be assessed is required to implement special assessments.

3. Revenue Bonds

These bonds are payable from charges made for services provided. Such bonds have advantages when agencies lack other means of raising capital and can be used to finance projects which extend beyond normal agency boundaries. The success of revenue bonds depends upon economic justification for the project, reputation of the agency, methods of billing and collection, rate structures, provisions for rate increases, financial management policy, reserve funds and forecast of net revenues.

4. Pay-As-You-Go Financing

This method entails gathering sufficient funding prior to and during construction. Funds may be gathered through a system of increased user charges and/or connection fees. Advantages of this method are elimination of interest cost and voter authorization. The principal disadvantage is significantly higher charges during the period when funds are being collected.

5. Revenue Programs and Rate Setting

Various methods are in use to obtain the revenue needed to operate, maintain, replace, extend, and enlarge wastewater systems. The principal methods include the levy of ad valorem taxes, service charges, connection charges, and combinations of these. Each agency's revenue program and rate settings reflect the needs of an individual community and its local policy. No single revenue generating program can be considered ideal for every situation. Alternatives should be examined in terms of equity among user groups, promotion of water conservation, and implementation and updating requirements.

Because the aim is to achieve fairness to all users and beneficiaries — domestic, commercial and industrial — it becomes necessary to consider the question of fairness as a whole and not simply from the standpoint of any one class. The procedures for allocating user costs are easier to understand in specific examples than in the general terms considered here. Examples are contained in "Financing and Charges for Wastewater Systems," a Joint Committee Report of APWA, ASCE and WPCF (3).

a. Ad Valorem Taxes

Taxes on real estate may be used as a primary revenue source for wastewater service in the event the agency has not accepted a federal grant and, therefore, is not subject to restrictions under the Clean Water Act or by taxing limitations. Real estate values and real estate taxes often reflect ability to pay, which may be a primary criterion in reducing hardship for some citizens. On the other hand, the value of real estate may have little relationship to the cost of service which is provided to individual users.

There is often a preference for payment of service as a tax rather than as a service bill because the tax may be deductible from federal and state income taxes, whereas a service charge is not.

The strongest objection to ad valorem taxation as a method of cost recovery comes from those who believe that benefits are strictly proportional to quantity of wastewater or a combination of quantity and waste characteristics.

The simplicity of an ad valorem tax system is a marked advantage. It requires no wastewater measurement or sampling program for charge purposes. The required accounting and billing work is minimal.

b. Service Charges

Service charges provide financial support of wastewater systems based on some measure of actual physical use of the system. Measures of use include: volume of wastewater; volume of wastewater plus quantity of pollutant matter; number or size of sewer connections; type of property, such as residential, commercial or manufacturing; number and type of plumbing fixtures, water using devices, or rooms; uniform rates per connection; and percent of water charge.

c. Connection Charges

A one-time charge at the time a user connects to a wastewater system is used to generate revenue. Charges vary from a nominal inspection fee to a full prorated share of the cost of the entire wastewater system. In some cases connection fees reflect costs to provide new capacity in the wastewater system. Connection fees are not generally used for operation and maintenance, but are used for purposes such as payment of debt service and financing wastewater system expansion.

d. Combination Taxes and Service Charges

Combination systems have often resulted from an original ad valorem tax system, when additional sources of revenue were required and taxes were limited, unpopular, or precluded from use to fund operation and maintenance costs. The Clean Water Act requires that grantees implement a system of service charges based on usage to pay for operation and maintenance costs.

The combination system can result in payment, by users and non-users who benefit from facilities, of an amount which is approximately in proportion to the cost of providing the use and the benefits of the wastewater system.

J. SAFETY

The goal in sanitary sewer projects is to eliminate unsafe conditions and unsafe acts. To be effective, desire and enthusiasm for safety must be encouraged and supported at all levels of employment in the owner organization. Among the numerous laws, rules, and regulations which govern safety is the Occupational Safety and Health Act of 1970 (OSHA).

1. Investigations

(a) Investigations or surveys, such as an EPA Sanitary Sewer Evaluation Survey (SSES), should be conducted with safety as a paramount consideration in accordance with regulations for equipment, training and size of sewer entry crew.

(b) The engineer must provide a work environment for his workers which recognizes hazards likely to cause death or serious physical harm. This can be accomplished with the utilization of proper safety equipment and procedures.

2. Design

(a) Safety factors are to be considered by the engineer for reducing the ultimate strength of a material to a working strength. The factor of safety will vary, depending on the type of material and its use. (A comprehensive discussion of design considerations is presented in Chapter 9.)

(b) The engineer must learn about safety practices and regulations such as OSHA Standards and how their requirements may affect the design of the sanitary projects.

(c) Sanitary sewers should be separated from gas and water mains and other buried utilities. Many states have regulations which set minimum horizontal and vertical separation requirements.

(d) Ventilation should provide for air to enter the sanitary sewer system and provide for the escape of gases. Currently, the trend is to utilize solid or watertight manhole covers in streets and water courses. With the decrease in ventilation through holes in manhole covers, inconspicuous vents or forced draft may be needed to provide adequate ventilation.

3. Construction

(a) The construction contract documents should require the contractor to adhere to all laws and regulations which bear on the project construction and to be responsible for safety at the construction site.

(b) The OSHA Standards require the employer (contractor) to provide employees with a safe and healthful place of employment. Sections of the OSHA Standards state that it shall be the responsibility of the employer to initiate and maintain such programs as may be necessary to provide safe conditions, and that frequent and regular inspections be made by competent persons designated by the employers.

(c) The engineer should specifically instruct his own field personnel to follow safety precautions while visiting the construction site. Field personnel should be issued hardhats and eye goggles, or other appropriate protective equipment for use while visiting the site.

(d) The terms of the contract documents entered into for a construction project play an important role in determining whether or not an engineer has any duty in regard to the safety of the contractor's employees. Litigation by injured employees of contractors naming the engineer as a party to a suit are occurring. The contract documents must provide that the responsibility for construction site safety is on the contractor. The engineer must, nevertheless, take appropriate action if a potential safety hazard is observed.

4. Operation and Maintenance

(a) A safety program is necessary for the operation and maintenance of a sanitary sewer system. Safety is the responsibility of every individual, not only for personal protection, but also for the protection of fellow employees.

(b) Safety equipment such as traffic control devices, safety harnesses, tools for manhole cover removal, gas detectors, blowers with duct discharge for positive displacement of manhole atmosphere, and rubberized cloth gloves must be available for protection where needed.

K. NATIONAL ENVIRONMENTAL POLICY ACT OF 1969

The National Environmental Policy Act of 1969 (NEPA) establishes that Congress is interested in restoring and maintaining environmental quality. It directs all Federal agencies to identify and develop methods and procedures to insure that environmental factors be given appropriate consideration in decision making along with economic and technical considerations. NEPA applies also to any project implemented by federal funds. A number of states have adopted legislation which parallels NEPA for state and local activities.

If the nature of the proposed project and its impact does not obviously reflect the need for the preparation of an Environmental Impact Statement (EIS), an initial environmental assessment reviews the anticipated effects of the proposed action. A conclusion is drawn as to whether the proposed action would significantly and adversely affect the environment, and, therefore, require the preparation of an EIS. If there is no significant adverse impact on the environment, then the assessment normally serves as the basis for a "negative declaration."

Wastewater projects can affect the types and intensities of land use by providing facilities to accommodate new development. The location of new or expanded facilities affects the location of new development. With limited public funds, a decision to finance a specific facility limits funds for other facilities locally or in other areas and for similar or other purposes. A compatible relationship needs to exist between the potential growth-accommodating effects of facilities and the ability of a region to support additional growth.

The assessment should address relative impacts of population growth on natural resources, such as air and water, and on visual conditions, community characteristics, archaeology, and other cultural values. It should clearly illustrate the range of impacts resulting from various growth levels, ranging from "no-growth" or "no project," to buildout.

Socioeconomic effects, employment, availability of housing, and the costs of providing projected populations with other public services and utilities, should be assessed. Those long-term environmental goals which conflict with economic needs must also be identified in assessment of secondary impacts.

An environmental assessment is an environmental report prepared expressly to determine whether a proposed action requires the preparation of an environmental impact statement. The environmental assessment may be based on procedures such as checklists, matrices, networks, overlays, and specific studies. The environmental assessment should:
- describe the proposed action;
- describe the environment to be affected;
- identify all relevant environmental impact areas;
- evaluate the potential environmental impacts;
- identify adverse impacts that cannot be avoided should the action be implemented;
- identify irreversible and irretrievable commitments of resources;
- discuss the relationship between local short-term uses of man's environment and long-term productivity;
- identify conflicts with state, regional, or local plans and programs;
- evaluate alternatives to the proposed action;
- discuss any existing controversy regarding the action.

It is important to quantify impacts where possible to permit a clear picture of the issues for discussion and decision-making purposes.

L. MEASUREMENT UNITS

In the United States, sanitary engineering technology has been based on measurements expressed in the foot-pound-second system previously prevalent in most English speaking countries (4). Metrication throughout the world is proceeding at varied paces; however, standardized measuring units of "Le Systeme International D'Unites" (SI) are becoming the rule rather than the exception. This manual, where practical, presents both metric and conventional systems.

The SI makes use of only seven base units which are divided into three classes:

(1) *Base Units.* These are seven well-defined units that are dimensionally

Table 1-1 Applicable SI Base Units

Quantity (1)	SI Unit (2)	Symbol (3)	Definition (4)
Base Units:			
Length	meter	m	The meter is the length equal to 1, 650, 763.73 wavelengths in vacuum of the radiation corresponding to the transition between the levels $2p_{10}$ and $5d_5$ of the krypton-86 atom.
Mass	kilogram	kg	The kilogram is equal to the mass of the international prototype of the kilogram.
Time	second	s	The second is the duration of 9, 192, 631, 770 periods of the radiation corresponding to the transition between the two hyperfine levels of the ground.
Supplementary Units:			
Plane Angle	radian	rad	The radian is the unit of measure of a plane angle with its vertex at the center of a circle and subtended by an arc equal in length to the radius.
Solid Angle	steradian	sr	The steradian is the unit of measure of a solid angle with its vertex at the center of a sphere and enclosing an area of the spherical surface equal to that of a square with sides equal in length to the radius.

Table 1-2 Applicable SI-Derived Units Expressed in Terms of Base Units

Quantity (1)	SI Unit (2)	SI Symbol (3)
Acceleration	Meters per second squared	m/s^2
Area	Square meters	m^2
Density	Kilograms per cubic meter	kg/m^3
Specific volume	Cubic meters per kilogram	m^3/kg
Velocity	Meters per second	m/s
Viscosity, kinematic	Square meters per second	m^2/s
Volume	Cubic meters	m^3

independent: meter (m), kilogram (kg), second (s), ampere (A), candela (cd), Kelvin (K), and mole (mol). Base unit definitions applicable to sewer design are given in Table 1-1.

(2) *Derived Units.* These are expressed algebraically in terms of base units. Several derived units have been given special names and symbols that may themselves be used to express other derived units in a simpler way than in terms of base units. Some of the derived units applicable to sewer design

Table 1-3 Applicable SI-Derived Units with Special Names

Quantity (1)	SI Unit (2)	Symbol (3)	Expression in terms of other units (4)	Expression in terms of SI base units (5)	Definition (6)
Force	newton	N		$m \cdot kg \cdot s^{-2}$	The newton is that force which, when applied to a body having a mass of one kilogram, gives it an acceleration of one meter per second per second.
Pressure	pascal	Pa	N/m²	$m^{-1} \cdot kg \cdot s^{-2}$	The pascal is the pressure or stress of one newton per square meter.

are shown in Tables 1-2 and 1-3.

(3) *Supplementary Units.* These are two units that are defined neither as base units nor as derived units. The units, radian (rad) and steradian (sr), are shown in Table 1.1.

Even though the selection of the meter-kilogram-second (mks) system yields coherent units more easily than those used, for example, in the centimeter-gram-second (cgs) system, not all of the units are convenient for all applications. Consequently, provision is made for multiples and submultiples of the base unit.

M. REFERENCES

1. *Glossary–Water and Wastewater Engineering.* APHA, AWWA, WPCF, ASCE, New York, N.Y., 1981.
2. *Regulation of Sewer Use,* Manual of Practice No. 3, Water Pollution Control Federation, Washington, D.C., 1975.
3. "Financing and Charges for Wastewater Systems," a Joint Committee Report of APWA, ASCE, and WPCF, New York, N.Y., 1973.
4. *Units of Expression for Wastewater Treatment,* Manual of Practice No. 6, Water Pollution Control Federation, Washington, D.C., 1976.

CHAPTER 2

SURVEYS AND INVESTIGATIONS

A. INTRODUCTION

Surveys and investigations produce the basic data needed for the successful conception and/or development of a sanitary sewer design project. The fundamental importance of any survey or investigation requires that it be carried out competently and thoroughly if an effective project is to result.

The term "survey" as used in this Manual refers to the process of collecting and compiling information necessary to develop any given phase of a project. In one sense it may include observations relating to general conditions affecting a project, such as historical, political, physical, environmental, and fiscal matters. In another sense, a survey may comprise the precise instrument measurements necessary for the engineering design.

The term "investigation" often is used interchangeably with "survey." Its use in this Manual, however, usually refers to the assimilation and analysis of the data produced by surveys to arrive at policy and engineering decisions.

Surveys and investigations for the preliminary phase of a project are broad in nature, with emphasis on covering all factors relating to a project and determining the relative importance of each. Surveys and investigations for design and construction phases are more precise and detailed, usually being limited by the scope of the project.

Methods of conducting a survey vary widely, depending on the phase of development under consideration and the objectives. Proper surveys require broad knowledge of the particular field and an understanding of the problems to be solved in the phase of the project for which the survey is being conducted. Knowledge of the various aspects of sanitary sewer design as set forth in other chapters of this Manual will lead to recognition of the specific information needed from a survey for any given project. The objectives of the survey for the several project phases and the type of information required for each phase are discussed in this chapter.

B. TYPES OF INFORMATION REQUIRED

Several different kinds of information, applicable in varying degrees to the different project phases, may be collected during the course of surveys for a typical project. These include:

1. Physical

(a) Topography, surface and subsurface conditions, details of paving to be disturbed, underground utilities and structures, subsoil conditions, water table elevations and traffic control needs.

(b) Locations of streets, alleys, or unusual obstructions; required rights-of-way; and all similar data necessary to define physical features of a proposed sanitary sewer project including preliminary horizontal and vertical alignment.

(c) Details of the existing sanitary sewer system, to which a proposed sanitary sewer may connect.

(d) Pertinent information relative to possible future extension of the proposed project by annexation or service agreements with adjacent communities or areas.

(e) Locations of historical and archaeological sites, and of people, plant and animal communities or any other environmentally sensitive area.

2. Developmental

(a) Population trends and density in area to be served.

(b) Type of development, i.e., residential, commercial, or industrial.

(c) Historical and experience data relating to existing facilities which may affect proposed sanitary sewers.

(d) Comprehensive regional master plans of other agencies, especially Clean Water Act (as amended) Section 208 Areawide Plans and Section 201 Facility Plans.

(e) Location of future roads, airports, parks, industrial areas, etc., which may affect the routing and location of sewers.

3. Political

(a) Present political boundaries and probability of annexation of adjacent areas.

(b) Possible service agreements with adjacent communities; feasibility of multi-municipal or regional system.

(c) Existence and enforcement of industrial waste ordinances regarding pretreatment or limitations on the concentrations of damaging substances.

(d) Requirements for new waste ordinances to achieve desired results.

4. Sanitary

(a) Quantity and strength of municipal wastewater to be transported.

(b) Water use data and flow gagings, where appropriate, to establish flow rates from existing similar areas.

(c) Capacity and condition of existing sanitary sewer system.

(d) Other pertinent data necessary to establish the required design criteria and capacity for the given project, including infiltration/inflow requirements and ordinances.

(e) Effectiveness and adequacy of present political subdivision to undertake the project; desirability of a new organization to sponsor same.

5. Financial

(a) Information relative to existing authorization or policies, obligations and commitments bearing on financing of proposed sanitary sewer.

(b) Amounts, retirement schedule, and refinancing penalty of outstanding bonds and unobligated bonding capacity available for proposed project.

(c) Availability of federal or state assistance through grants or loans.

(d) Taxable valuation, existing tax levies, and any limits affecting proposed project.

(e) Schedule and method of existing sanitary sewer service rates and revenues generated.

(f) Property plots as required for sanitary sewer assessments and special methods of assigning assessments.

(g) History of local construction factors and operating costs, and conditions affecting cost.

(h) All similar data necessary to establish a feasible financing program for the proposed project.

C. SOURCES OF INFORMATION

Possible sources of information sought by surveys for sanitary sewer projects include:

1. Physical

(a) Existing maps and sewer system plans, including United States Geological Survey topographic maps, city plots and topographic maps, state highway plans and maps, tax maps, local utility records and plans.
(b) Aerial photographs.
(c) Instrument surveys, including approximate surveys by such devices as hand level and aneroid barometer which may be useful for preliminary work.
(d) Photographs of complex surface detail to supplement instrument surveys, and photographs to show detail of existing sewer systems.
(e) Borings and test pits, either by hand or by machine, for determining subsurface soil and water conditions, also sounding-rod probing for underground structure location and an indication of soil conditions.
(f) Local 208 Agency, Historical Preservation Office, Archaeological Society, Department of Natural Resources, Soil Conservation Service or Local Universities to identify any environmentally sensitive areas.

2. Developmental

(a) Census reports.
(b) Planning and zoning reports and maps.
(c) General field and/or aerial photo examination to note type, degree, and density of development.
(d) Criteria of regulatory agencies having jurisdiction over the project.
(e) Engineering reports or studies of related projects in the area.

3. Political

(a) Enabling or authorizing legislation.
(b) Municipal and state laws.
(c) Conferences with owner and other officials.
(d) Comprehensive plans established by planning agencies.
(e) Local and area meeting reports and minutes.

4. Sanitary

(a) Canvass of significant industry to determine type and amount of waste.
(b) Flow gagings and sampling in existing sanitary sewers to establish flow characteristics from similar areas.
(c) Records of water pumpage and water sales.
(d) Design basis and operational characteristics of existing sanitary sewers from system records.
(e) Federal, State, and Local requirements and ordinances for infiltration/inflow.
(f) Records of existing treatment plant influent characteristics.

5. Financial

(a) Pertinent records of the owner's fiscal officer.
(b) Auditor's or treasurer's records relating to tax levies.

(c) Operating statements and reports of income and expense for the sanitary sewer, water, and other utility departments.

(d) Ordinances or laws and bond indenture governing outstanding bonds, and procedures for financing and contracting the proposed project.

(e) Assessment plots and schedules for prior projects to show methods in use in the locality.

(f) Tax maps showing subdivision and ownership of property to be affected by special assessments.

D. SURVEYS FOR DIFFERENT PROJECT PHASES

The objectives of the typical and differing phases of project development must be understood if the survey to be conducted is to be meaningful. Typical phases of project development are discussed in Chapter 1. A discussion of the objectives and the nature of surveys for each phase follows.

1. Preliminary Surveys

This phase of development is concerned with the broad aspects of the project, including such things as required capacity, basic arrangement and size, probable cost, environmental assessment, and methods of financing. Accordingly, information is required in sufficient detail to show general physical features affecting layout and general design. The scope as well as the feasibility and environmental aspects of the project development must be considered at this time. When a connection to an existing sanitary sewer system is proposed, the preliminary survey must contain flow data for use in establishing the design capacity and sanitary sewer layout.

Extreme precision and detail are neither necessary nor desirable in this phase, but all data obtained must be reliable. The type and extent of the information needed for any given project usually will become apparent as the work progresses, and they may vary widely depending on the size and complexity of the project. Since preliminary surveys are used to develop the information on which the estimates of the engineering report are based, these must be thoroughly and competently prosecuted. Sufficient allowance must be made for items affecting the total cost of the project, such as trends in construction costs, pavement removal and replacement practices, backfill methods, sheeting, well pointing and unusual quantities of difficult excavation. Otherwise, costs will be underestimated.

Occasionally, photogrammetric methods may be advantageous in obtaining a portion of the data needed in making a preliminary survey.

2. Design Surveys

Surveys for this phase form the basis for engineering design as well as for the preparation of plans and specifications. Design surveys are concerned primarily with obtaining physical and sanitary wastewater flow data rather than developmental, environmental or financial information. In contrast with the preliminary surveys, design surveys must contain all the detail and precision the design engineer needs to correlate his design and the resulting construction plans with actual field conditions. Design surveys involve the use of surveying instruments in establishing the accurate location of pertinent topographic features. Photogrammetric methods also may be used in obtaining this information.

Presumably, the preliminary phase will have established the general extent of the project so that the area to be covered by the design survey can be defined. However, further definition of location within the general area may be required during the course of the design survey. The design survey also may extend to some degree beyond the proposed construction limits in order that possible future expansion may be facilitated.

Obviously, accurate surveys are required to produce accurate designs and plans. Vertical control usually is established by setting benchmarks throughout the project, the elevations of which have been checked by level circuits to within 3 mm (0.01 ft). Although property and street lines are often used for horizontal control, in large projects control traverses or coordinate systems may be desirable. A special problem frequently encountered in design surveys is the nature and extent of existing underground utilities and structures which must be cleared or displaced by the new sanitary sewer. Such information, insofar as practicable, must be obtained during the design survey to establish rights-of-way, minimize utility relocation costs, obtain lower construction bids, and prevent changing or realignment after the commencement of the project. Where the accurate location of important substructures cannot be ascertained by other means, and conflict is possible, excavation to determine location, elevation, and detail at the point of crossing is warranted. Such substructure information may be determined by the owner, engineer or contractor prior to sewer construction with alignment or invert elevations revised accordingly.

3. Construction Surveys

Surveys for this phase are concerned almost exclusively with physical aspects. Construction surveys are required to establish control for line and grade, to check conformity of construction, and to establish settlement levels on existing structures immediately adjacent to the construction where necessary.

E. INVESTIGATIONS

Investigations may take many forms but always are directed toward determining the most feasible, practical and economical methods of achieving a desired result. On small sanitary sewer projects, these may involve no more than an on-the-spot decision to use conventional minimum standards for a simple gravity flow extension to an existing sewer system. Larger projects, on the other hand, may have several alternatives, all of which must be considered. Projects involving relief of existing sewer systems, for example, usually require extensive studies before the design flow capacity and the method of correction can be ascertained.

The following questions are typical of those to be resolved by investigation:

(a) What is the extent of area to be served and what is the pattern of present and future land use? Has an area zoning plan been adopted? How does the area relate to a regional sanitary sewer plan?

(b) What general arrangement of the system will best fill the need? What easement or rights-of-way are required for this arrangement?

(c) What part of the wastewater flow shall be intercepted for treatment from an existing sewer system?

(d) Are there combined sewers in the system? How will flow from combined sewers be handled?

(e) What are the estimated present and future wastewater flows?

(f) Shall sanitary sewers all discharge to one point for treatment or shall treatment be provided at more than one location?

(g) How will requirements of other agencies (state and county highway departments, railroads) dictate specific locations for crossing, rights-of-way, installation, and details of materials or construction?

(h) If a regional sewer system is anticipated, what are the long-term environmental effects of exporting wastes from a given groundwater basin to another?

CHAPTER 3

QUANTITY OF WASTEWATER

A. INTRODUCTION

Sanitary sewers are constructed primarily to transport the wastewater of a community to a point of treatment or ultimate disposal. Wastewater may be characterized as domestic, commercial or industrial in origin. Other extraneous waters, such as infiltration or inflow, should be excluded insofar as practicable, and connection of roof, yard, and foundation drains to the sanitary sewers should be prohibited through local legislation. Building construction and grading practices which permit surface water to enter basements or crawl spaces and subsequently drain to the sanitary sewer through illegal connections similarly should be prohibited.

The sanitary sewer capacity must be determined from careful analysis of the present and probable future quantities of domestic, commercial and industrial wastewaters, as well as anticipated groundwater infiltration and extraneous inflow entering through specific connections such as basement drains or similar inflow sources. Sanitary sewers are ordinarily sized on the basis of design flows made up of peak wastewater flow and infiltration. However, in some sewers peak discharges or surcharges resulting from infiltration/inflow may be greatly in excess of ordinary peak design flows.

The terms infiltration/inflow are defined as:

Infiltration—The total extraneous flow entering a sewer system or portions thereof, excluding sanitary sewage, because of poor construction, corrosion of the pipe from the inside or outside, ground movement or structural failure through joints, porous walls or breaks.

Inflow—The extraneous flow which enters a sanitary sewer from sources other than infiltration, such as roof leaders, basement drains, land drains, and manhole covers. Inflow, in short, is man-made and intentional.

B. DESIGN PERIOD

Every political subdivision having a sanitary sewer system and anticipating future growth either in terms of population or commercial and industrial expansion should have a long-range plan for the installation of sanitary sewers within its tributary area. The plan should be flexible, current, and coordinated with the facilities plan or zoning ordinance for the community and relevant adjacent areas. For any given drainage area the plan should contain, with appropriate modifications, most of the engineering data discussed in this chapter. A long-range plan permits orderly expansion of facilities based on sound engineering without resort to costly "crisis" action.

The length of time that the capacity of sanitary sewers will be adequate is referred to as the design period. It must be established prior to the design of the sanitary sewer.

Sanitary trunk sewers, interceptors, and outfalls are generally designed for the projected peak flow rate expected during a 50-year period. For large sanitary sewers, past and future trends in population and water use, and existing wastewater flows must be studied before a design period can be selected. In the case of sanitary trunk sewers serving relatively undeveloped lands adjacent to metropolitan areas, it may not be feasible to design and construct initial facilities for more than a limited future development period. Nevertheless, easements and rights-of-way for future facilities should be secured during the original construction program or as far in advance of development of the area as possible.

Curent trends, insofar as the U.S. Environmental Protection Agency (EPA) is concerned, are to limit the design period to between 10 and 15 years. In any case it would be prudent to design the facilities to at least outlast the bond issue or funding for the project.

C. DESIGN FLOWS

Once the design period has been determined, consideration must be given to the quantity of wastewater to be transported. Because the flow is largely a fuction of population served, population density, and water consumption, sanitary sewers should be designed for peak flow rates corresponding to the population at saturation density as set forth in the community's facility plan or as otherwise predicted for the design period. In systems relatively free of unwanted infiltration/inflow, anticipated maximum water use rates and population density may be used as a guide in determining maximum wastewater flow. Nonetheless, a critical review is due any apparent relationship between water use and wastewater discharged. Where infiltration/inflow is a major factor, the number of structures or extent of area served or to be served may be more significant than the population or maximum water use.

Design flows for sanitary sewer systems can be separated into two categories: Those in which wastewater is the major flow and only minor allowances need to be made for infiltration/inflow; and those in which infiltration/inflow dominate. In the latter case, the wastewater contribution may be a minor factor in determining the peak flow rate. Clearly, the connection of roof, yard, areaway, and foundation drains to sanitary sewers must be prohibited if this second category is to be avoided. Yet it must be recognized that established usage, as well as topographical and political considerations, may make all or part of this approach impractical as a corrective measure.

The designer, however, must determine which of these categories, or what combination of them, exists or may be anticipated. For example, infiltration/inflow in a relatively new sanitary sewer system in one midwestern suburban community (1) was found to be as high as 0.08 m^3/min/ha (0.02 cfs/acre) or in excess of 4.94 m^3/day/cap (1,300 gpd/cap). Average dry-weather flows, on the other hand, were less than 0.26 m^3/day/cap (70 gpd/cap). A study carried out in another community by other investigators recorded peak flows of 0.18 m^3/min/ha (0.042 cfs/acre) and 0.14 m^3/min/ha (0.003 cfs/acre), respectively, in two suburban subdivisions. Under conditions this severe it may not be practicable to convey wet-weather peak flow to central treatment plants; other means for wastewater handling may be required (2).

Construction costs of sanitary sewers designed to accommodate extreme rates of infiltration/inflow are greater than for those carrying only sanitary wastewater. Efforts should be made to locate , quantify and determine the cost effectiveness of corrective action to reduce or eliminate sources of extreme infiltration/inflow prior to design of oversized facilities.

Flows important to the design of sanitary sewers and treatment facilities, particularly where wastewater contributions govern, include daily minimum and maximum, daily mean or average, and peak flows. *Peak flow* may be defined as the greatest volume of influent to a treatment plant within a given time period. It should be recognized that flows normally will vary greatly during the design period. *Mean daily flow* of wastewater including groundwater in an existing sewer system may be derived from an analysis of a full year's metering data, or from an infiltration/inflow analysis. Flow information obtained at the plant by metering devices aids in estimating minimum, maximum, and peak flows where they cannot be measured in the system. *Daily minimum and maximum* discharges for the initial and final years of the design period are of value in determining treatment plant capacities. Peak flows estimated for the end of the design period determine the hydraulic capacities of sanitary sewers and some treatment plant conduits and pumpage. Minimum flows are related directly to the design of sanitary sewers to insure proper transport of sediment during low velocities. Minimum flow estimates facilitate consideration of measures required to protect sanitary sewers against sulfide-related corrosion as discussed in Chapter 4.

A sanitary sewer has two main functions: To carry the peak discharge for which it is designed, and to transport suspended materials to prevent deposition in the sanitary sewer. It is essential, therefore, that sanitary sewers have adequate capacity for the peak flow and that they function at minimum flows without operational problems (16).

1. Population Estimates

In estimating the sanitary wastewater fraction of the average daily flow, it is customary to multiply the estimated future tributary population by the estimated per capita wastewater production. Obviously, the accuracy of the population estimate is important in this computation. Efforts should be made to assemble and study as much population information as possible from various corroborating sources (see Chapter 2). The accuracy of long range population forecasts, however, is limited due to numerous economic and social variables that influence population growth trends. This limited accuracy should be considered in forecasting wastewater flow rates. Care should be exercised where seasonal or transient populations are included. For use in arriving at annual average flows, seasonal population estimates can be converted to equivalent full-time residents by using the following multipliers of the population fraction that is considered seasonal:

Day-use visitor, 0.1 to 0.2
Seasonal visitor, 0.5 to 0.8

City planning studies or Clean Water Act (as amended) Section 208, Areawide Waste Treatment Management Planning information may be available in which population estimates have been developed presumably with great care and in considerable detail. As always, verification is prudent.

Future population trends depend on many factors. Known factors include location of transportation facilities for workers, raw materials, and manufactured products; possible expansion of present industries; availability of sites for residential, commercial, or industrial development; civic interest in community growth; availability of other utility services at reasonable rates; and real estate values. Nevertheless, population estimates based on all known factors can be upset by extraordinary events such as the discovery of some new natural resource in the vicinity or the decision by a large manufacturer to locate in the community.

Applicable methods used for population projections have included:

(a) Arithmetical increase per year or per decade;
(b) Uniform percentage rate of growth based on recent census periods;
(c) Decreasing percentage rate of increase;
(d) Graphical comparison with the growth of other similar but larger cities;
(e) Graphical extension of the curve of past growth; and
(f) Verhulst's theory (logistic trend method) (3).

An example of a suitable method of population prediction is illustrated in Fig. 3-1. Probable future growth is indicated by two limiting lines, as shown, because of the difficulty of predicting future population with any great degree of accuracy.

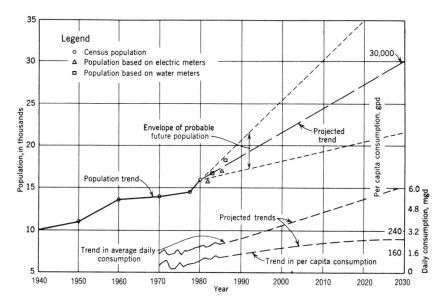

Fig. 3-1. Population and water consumption trends (gpd/cap × 3.8 = L/day/cap; mgd × 3.8 = m³/day).

Demographers tend to use more sophisticated methods of population projections, such as ratio and correlation, component analysis, and employment forecast which may be useful for detailed studies of the area (4).

2. Land Use

City or urban area planning studies and areawide wastewater treatment managment planning reports, where available, are a valuable guide in determining future land use. Present land use as given by zoning regulations and the history of population migrations are also useful guides in studying population changes and distribution within a community. The trend toward decentralization of business and industry, or the result of the "energy crunch" as a reversing condition, has and will continue to have a complicating influence on population predictions.

3. Tributary Areas

Tributary areas for sanitary sewer design may be limited by natural topography, political boundaries, or economic factors. The sanitary sewer design for small drainage areas normally should provide sufficient capacity to serve the entire area (5).

Political boundaries sometimes will determine the extent of a project if legal restrictions prevent the financing of construction beyond the limits of the municipality or district. Nonetheless, sanitary sewer capacities should be based on total tributary area, if possible, since political boundaries and legal restrictions may change based on needs.

The economics of project financing also may place limits on the design; that is, the design may be restricted to the political boundaries or even a smaller area if intrapolitical financing arrangements cannot be made or if the forecasted population growth rate is too low to justify extending the sewer system to a larger portion of the tributary area.

4. Per Capita Wastewater Flow

Dry weather flow quantities usually are less than the per capita water consumption because water is lost through leakage, lawn irrigation, swimming pools, etc. In arid regions the mean wastewater flow may be as little as 40% of the average water consumption. In estimating probable future per capita wastewater contributions consideration should be given to changes in water use habits and household appliances, such as garbage grinders, dishwashers, and automatic washers, as well as to the growing conservation ethic and water conservation requirements. Average daily per capita domestic wastewater flows used for design purposes in a number of geographic locations are found in Table 3-1. Most of these flows include only nominal allowance for infiltration/inflow, but there are exceptions. Figs. 3-2 and 3-3 are design curves used in Austin and Dallas, Texas. Both show that as the service area increases, the per acre contribution decreases.

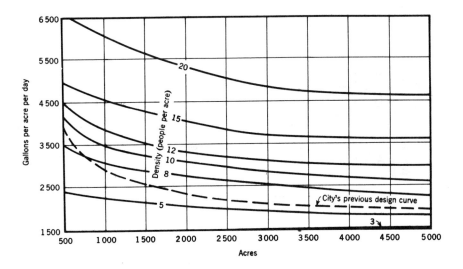

Fig. 3-2. Design flows used in Austin, TX. (gpd/acre × 0.00935 = m³/day/ha; acre × 0.405 = ha).

Table 3-1 Typical Design Flows

City (1)	Year and Source of Data (2)	Average Rate of Water Consumption (gpd/cap) (3)	Population Served (1,000's) (4)	Average Sewage Flow (gpd/cap) (5)	Sewer Design Basis (gpd/cap) (6)	Remarks (7)
Baltimore, Md.	....	160	1,300	100	135 × factor	Factor 4 to 2
Berkeley, Ca.	....	76	113	60	92	
Boston, Mass.	1980	215	640	264	100*	*Includes infiltration. Multiply by 3 when sewer is flowing full.
Cranston, R.I.	1980	120	74	152	150	
Des Moines, Iowa	1980	132	230	100	100 × Factor*	*Factor $= \dfrac{18 + \sqrt{P}}{4 + \sqrt{P}}$ P = population in thousands
Detroit, Mich.	1964	229	3,500	196	259 324	For separate sewers For combined sewers (sewer design basis is for peak flows and include allowances for infiltration and ordinary public, commercial, and industrial flow).
Grand Rapids, Mich.	1980	178*	240	190*	100 × factor**	*Including commercial and industrial services. **Factor 2.5 to 3.5 for sewers 24 in. diam and less, 2 to 3 for larger sewers.
Greenville Co., S.C.	1959	110	200	150	300	Service area includes City of Greenville. Sewers 24 in. diam and less designed to flow one half full at 300 gpd/cap; sewers larger than 24 in. designed to have 1-ft freeboard.

QUANTITY OF WASTEWATER

Table 3-1 Typical Design Flows (Cont.)

City (1)	Year and Source of Data (2)	Average Rate of Water Consumption (gpd/cap) (3)	Population Served (1,000's) (4)	Average Sewage Flow (gpd/cap) (5)	Sewer Design Basis (gpd/cap) (6)	Remarks (7)
Hagerstown, Md.	....	100	38	100	250	
Jefferson Co., Ala.	....	102	500	100	300	
Johnson Co., Ks.	1958					
Mission Township Main Sewer District		70	70	60	1,350	Most houses have basements with exterior foundation drains.
Indian Creek Main Sewer District		70	30	60	675	Most houses have basements with interior foundation drains.
Kansas City, Mo.	1980	180	560	160	0.01 cfs/acre 0.02 cfs/acre	For interceptors and trunk lines. For laterals and sub-mains. Many houses have basements. Foundation drains are not allowed to be connected to sanitary drains.
Lancaster Co., Neb.	1962	167	148	92	400	Serves City of Lincoln.
Las Vegas, Nev.	....	410	45	209	250	
Lincoln, Neb. (Lateral Dists.)	1964			60 60		For lateral sewers max flow by formula: Peak flow = 5 × avg flow ÷ (Pop. in $1,000's)^{0.2}$

Table 3-1 Typical Design Flows (Cont.)

City (1)	Year and Source of Data (2)	Average Rate of Water Consumption (gpd/cap) (3)	Population Served (1,000's) (4)	Average Sewage Flow (gpd/cap) (5)	Sewer Design Basis (gpd/cap) (6)	Remarks (7)
Little Rock, Ark.	....	50	100	50	100	
Los Angeles, Ca.	1979	180	2,816	85	*	*85 gpd residential multiplied by peak factor. See Fig. 3-6.
Los Angeles Co. San. Dist., Ca.	1979	197	3,650	82**		**82 gpd/cap. is the average residential flow. Average flow from all users, including industry, is 120 gpd/cap.
Monroe Co., N. Y.	1978	110	300*	120	300**	*Includes all suburban population. **For sanitary interceptor sewers (all figures include industrial and commercial water consumption and sewage generation).
Greater Peoria San. Dist., Ill.	1980	70	165	100	800 500	Laterals and sub-trunk sewers. Trunk sewers and interceptors.
Madison, Wis.	1979	60	170	48	100	Maximum hourly rate
Milwaukee, Wis.	1980	175	670	225	275*	*Plus additional factors for inflow/infiltration
Memphis, Tenn.	....	125	450	100	100	
Orlando, Fla.	1980	176	129	155*	250	*25% infiltration rate included.
Painesville, Oh.	1979	190	20	175	400	Includes infiltration.

Table 3-1 Typical Design Flows (Cont.)

City (1)	Year and Source of Data (2)	Average Rate of Water Consumption (gpd/cap) (3)	Population Served (1,000's) (4)	Average Sewage Flow (gpd/cap) (5)	Sewer Design Basis (gpd/cap) (6)	Remarks (7)
Rapid City, S.D.	1979	125	50	125	225	
Rochester, N.Y.	1978	140	400*	150	400** 300***	*Includes 150,000 suburban population. **For sanitary collector sewers. ***For sanitary interceptor sewers (all figures include industrial and commercial water consumption and sewage generation).
Santa Monica, Ca.	1980	137	93	95	100	
Shreveport, La.	1979	130	180	80	150	Sewer design 150 gpd/cap plus 600 gpd/acre infiltration. Sewers 24 in. diam and less designed to flow one half full; sewers larger than 24 in. designed to have 1-ft freeboard.
St. Joseph, Mo.	1960		85	125	450	Main sewers.
Wyoming, Mich.	1979	150	125	115*	195	*Calculated actual domestic sewage flow, not including infiltration or industrial flow.

NOTE: Gal x 3.785 = L; gpd/acre x 0.00935 = m³/day/ha; ft x 0.3 = m; in. x 2.54 = cm.

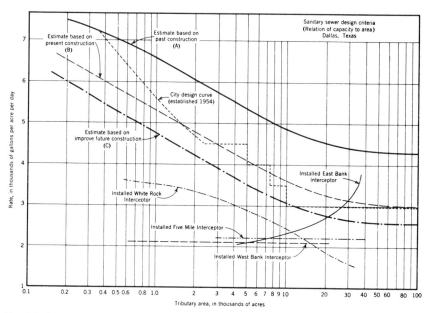

Fig. 3-3. Design flows used in Dallas, TX. (gpd/acre × 0.00935 = m³/day/ha; acre × 0.405 = ha).

5. Flow Estimates Based on Population and Flow Trends

Fig. 3-1 shows a typical plot of past census populations of a small city and estimates of the population made from the number of electric meters and water meters for each year subsequent to the last regular census year. The records of average daily and per capita water consumption, which may be used when measured wastewater flows are not available, are shown for the previous 15-year period.

Population and water consumption are predicted up to the year 2030. The zigzag nature of past population growth indicates the inappropriateness of using any mathematical curve to forecast future population. It is probable that the future population will, in the engineer's judgement, be included within the envelope shown by the dotted lines in Fig. 3-1. The upper line represents the maximum past rate of growth, and the lower line the minimum. In the illustration a projection trend has been adopted suggesting an estimated population of 30,000 at the end of the design period. Should the actual population growth follow the upper boundary of the envelope, the capacity of the sanitary system should be reached by about 2010. If the population growth should follow the lower boundary, the sanitary system should be adequate well beyond the year 2030.

The past trend in per capita water consumption in the example shows a slow increase. The future trend in average daily consumption has been estimated by multiplying the projected trend in per capita consumption by the projected trend in population. Wastewater flow rates may be estimated as a portion of the projected water consumption.

6. Distribution of Population

While population prediction provides information for design of future sewer use for a municipality, it does not usually provide specifics for smaller included areas. Land use and population density can provide insight into future growth patterns. Population density for the entire area or tributary area may be estimated by dividing the projected future population by the total area to be served by the sanitary sewers, with proper deductions for uninhabitable land and water areas. The saturation density of the community should also be estimated. The future design population density for a portion of the community normally should be greater than the existing average for the community as a whole up to the limit of the saturation density. It cannot be assumed that all small areas will reach saturation densities of population, yet there is no way of predicting which areas will. Accordingly, it is desirable that sanitary sewers serving small areas be designed on the basis of saturation densities. However, in order to avoid oversizing, interceptor design should be based on population projections for the entire area to be served.

Suburban saturation densities range from 5 or 7.5 cap/ha (2 or 3 cap/acre) for very large lots, to about 75 cap/ha (30 cap/acre) for small lots zoned for single-family residences. Urban densities may be as high as 125 cap/ha (50 cap/acre) for areas zoned for two-family residences on relatively small lots, and may exceed 2,500 cap/ha (1,000 cap/acre) for multistory apartment buildings. Estimates of saturation populations may vary greatly for any specific area.

7. Commercial Contributions

Wastewaters from commercial areas are usually estimated in terms of cubic meters per day per hectare (gallons per day per acre), based on a study of contributions from existing developed areas and comparable data from other cities. Sewer designs for commercial areas should be more generous due to the high costs and inconvenience of later replacement or provision of relief as well as the potential damage from surcharge. Smaller communities with no significant commercial development must allow for possible future commercial contributions. The allowance is usually considered to be included within the per capita flow estimates developed for domestic wastewater. Commercial areas in larger cities are often more complex than those in smaller communities. Because they may attract workers and customers from relatively long distances, commercial buildings may have more fixtures with greater usage, thus generating a greater wastewater flow per hectare (acre). Table 3-2 lists actual allowances for commercial contributions from areas in various cities. Note that the variation ranges from 42 to 1,500 m^3/day/ha (4,500 to more than 160,000 gpd/acre). Table 3-3 indicates ranges of average sanitary wastewater quantities from commerical areas. Coin-operated laundries and car, truck and bus washes may contribute substantial flows particularly where units are large or numerous. Actual flow measurements at apartment projects in Houston, Texas (6), in the early 1960s account for the results shown in Table 3-4. Additional data for apartments in Chicago are contained in Table 3-11.

Table 3-2 Sewer Capacity Allowances for Commercial and Industrial Areas

City (1)	Year and Source of Data (2)	Commercial Allowances (gpd/acre) (3)	Industrial Allowance (gpd/acre) (4)
Cincinnati, Oh.	1980	Case by case determination after consulation with the Directors of Sewers.	
Dallas, Tex.	1960	30,000 added to domestic rate for downtown; 60,000 for tunnel relief sewers.	
Grand Rapids, Mich.	1980	Offices, 40-50 gpd/cap; hotels, 400-500 gpd/room; hospitals, 200 gpd/bed; schools, 200-300 gpd/room.	
Hagerstown, Md.	.....	Hotels, 180-250 gpd/room; hospitals, 150 gpd/bed; schools, 120-150 gpd/room.	
Houston, Tex.	1960	Peak flows: Offices 0.36 gpd/sq ft; retail 0-20 gpd/sq ft; hotels, 0.93 gpd/sq ft.	
Las Vegas, Nev.	.....	Resort hotels, 310-525 gpd/room; schools, 15 gpd/cap.	
Los Angeles, Ca.	1980	Commercial, 100 gpd/1,000 sq ft gross floor area; hospitals, 500 gpd/bed (surgical), 85 gpd/bed (convalescent); schools, elementary or jr. high school — 10 gpd/student, high schools — 15 gpd/student; universities, 20 gpd/student. The above values give peak flow rates. Divide by 3.0 to obtain average flow rates.	15,500
Los Angeles Co. San. District, Ca.	1980	4,000 – 6,000	
Lincoln, Neb.	1962	7,000	
Milwaukee, Wis.	1980	240,000 (max) 25,800 (min)	
St. Joseph, Mo.	1962	64,000 (downtown) 25,800 (neighborhood)	
St. Louis, Mo.	1960	90,000 avg., 165,000 peak	
Santa Monica, Ca.	1980	Commercial, 9,700; hotels, 7,750	13,600
Toronto, Ont., Canada	1980	Analysis of actual water consumption in the commercial and industrial downtown area is approximately 20,000 gpd/acre.	

Note: gpd/acre × 0.00935 = m³/day/ha; gal × 3.785 = L; gpd/sq ft × 40.7 = L/d/m²

Table 3-3 Average Commercial Flows

Type of Establishment (1)	Avg Flow (gpd/cap) (2)
Stores, offices, and small businesses	12 to 25
Hotels	50 to 150
Motels	50 to 125
Drive-in theaters (3 persons per car)	8 to 10
Schools (no showers), 8-hr period	8 to 35
Schools (with showers), 8-hr period	17 to 25
Tourist and trailer camps	80 to 120
Recreational and summer camps	20 to 25

Note: Gal × 3.785 = L

Table 3-4 Measured Flow from Apartment Projects, Houston, Tex.

Stories per Structure (1)	Dwelling Units per Acre (2)	Flow (gpd/acre)		
		Average (3)	Peak Day (4)	Peak Hour (5)
2	16.6	3,430	3,690	14,900
2	20.3	4,340	5,880	15,000
15	80.3	8,720	10,200	20,000

Note: Acre × 0.405 = ha; gpd/acre × 0.00935 = m³/day/ha.

The transient population in large cities must be considered, as it may account for substantial flows. For example, consider the lower half of the Borough of Manhattan in New York City, where the population is constituted as follows: Residents, those who live in the area, regardless of where they work; workers, those who work in the area, regardless of where they reside; and transients, those who visit the area during the day for business, recreation, or other reasons.

Wastewater quantities for these several population categories are tabulated in Table 3-5 for a study area from West 14th Street to the Battery, between the Hudson River and Broadway.

8. Industrial Contributions (7)

Industrial wastewater quantities may vary from little more than the normal domestic rates to many times those rates (see Table 3-2). The type of industry to be served, the size of the industry, operational techniques and the method of on-site treatment of wastewater are important factors in estimating wastewater quantities. Furthermore, peak discharges may be the result of flows contributed over a short time frame (e.g., 10-hr working period). Peak discharge to the sanitary sewer sometimes can be reduced by the use of detention tanks or basins arranged to discharge at smaller rates over longer periods or to discharge only during hours when wastewater flows are small.

Table 3-5 Estimated Average Wastewater Flows in a Section of New York City

Contributor (1)	Population (2)	Water Consumption (gpd/cap) (3)	Wastewater Flow (mgd) (4)
Residents	67,000	200	8.55
Workers	409,000	30	8.56
Transients	392,000	15	3.80

Note: Gal × 3.785 = L; mgd × 3.785 = m³/day.

9. Institutional Contributions

Institutional wastewater contributions can generally be determined more accurately than commercial and industrial flows since pertinent data are more readily available. Records are generally available and the sanitary sewer use patterns may be well established, especially for governmental areas and buildings. Because similar institutions are often reasonably well related insofar as wastewater flows are concerned, a study of similar existing facilities may permit estimates to be made for new or unserved institutional facilities. However, care must be taken to adjust for variations in building and lot sizes when flows are expressed in area units.

Institutional flow measurements made at St. John's Hospital, St. Louis, Missouri, in the spring of 1960 are shown in Table 3-6. The hospital had 525 beds and a staff of 750 employed in three shifts. Rates of flow were recorded by instruments attached to each water meter supplying the hospital. It was estimated that approximately 93% of the water was returned to the sanitary sewers. Water usage for electric power and steam used by the hospital were furnished by others.

10. Cooling Water

All single-pass unpolluted cooling water used either for air conditioning or industrial processes should be kept out of sanitary sewers.

Table 3-6 Water Use at St. John's Hospital, St. Louis, Mo.

Item (1)	Water Use Rate (gpd/bed) (2)	Ratio of Designated Use to Average (3)
Average daily	300	1
Maximum day	375	1.25
Average 5-hr maximum	530	1.70
5-hr maximum	675	2.25
2-hr maximum	800	2.70

Note: Gal × 3.785 = L

Once-through cooling water for air conditioning equipment is used at the rate of 5.7 to 7.6 L/min (1.5 to 2 gpm) per ton of refrigeration. A 5-ton unit could use approximately 53 m³/day (14,000 gpd). It would not take many such units to seriously affect flow through a sanitary sewer. The application of demand water charges per ton of nonconserving air conditioning equipment and developments to eliminate or minimize water use suggest that the discharge of cooling water may become a less serious problem in the future.

11. Infiltration/Inflow Contributions (8)

Sanitary sewer design capacity must include an allowance for extraneous water components which inevitably become a part of the total flow. Proper design and construction will reduce the quantity of extraneous water entering the sanitary sewer as infiltration through cracked pipes and defective joints or as inflow through cross connections, faulty manholes, and submerged manhole covers. There is considerably less control over extraneous water entering from improper house connections, illegal drains, or other defects on private property. These aspects normally are controlled through enforcement of regulations or ordinances. Experience demonstrates that reasonable control through the inspection of house sewer construction, as well as the enforcement of the prohibition of roof and areaway drains or similar connections can be accomplished. But in some areas it may not be feasible to prohibit foundation drains which have been functioning for years.

It is important that the appropriate local agency formulate enforceable regulations which will assure that infiltration/inflow are kept within reasonable limits. The usual procedure is to balance the costs of control against the benefits obtained. Costs, for example, may include extra expenses for yard grading and building construction, or alternate means of disposal of foundation drainage (storm drainage). Financial benefits include the differential value of reduced sanitary sewer sizes and the present worth of all future excess pumping station and treatment plant capacities and operating costs. There is a point of balance for each sanitary sewer system beyond which the cost of further reduction of infiltration/inflow will not be offset by equal savings. The designer should locate this point as accurately as possible and design accordingly.

A more detailed discussion of infiltration/inflow and their control follows:

a. Inflow from Leakage into Manholes; Roof and Areaway Drainage; Surface Runoff into Basements and Crawl Spaces

Runoff from impervious areas represented by roofs and pavement should be kept out of sanitary sewers by enforced regulation. Tests made on manhole covers (9) submerged in only 25 mm (1 in.) of water indicate that the leakage rate per manhole may be from 1.3 to 4.7 L/sec (20 to 75 gpm) depending on the number and size of holes in the cover. Although such leakage could contribute quantities of stormwater several times the average sanitary flow, it can be minimized by using solid covers with half-depth pick holes or other proprietary devices. A few illegal roof drain connections also can overload smaller sanitary sewers. Rainfall of 25 mm/hr (1 in./hr) on 100m² (1,080 sq ft) of roof area, for example, would contribute water at about the rate of 0.7 L/sec (11 gpm). Direct entry of surface runoff into basements or drained crawl spaces through window wells, areaways, basement garages, or directly through foundation walls can result in flows of extreme magnitude. Regulations should be adopted and enforced to prevent or at least severly limit conditions of this sort. Since compliance with the regulations may increase the cost of yard grading and building construction, a determined and continuing resistance should be anticipated. The designer, therefore, must evaluate the situation and make allowances for such amounts of manhole leakage, roof water, and surface runoff as in his judgement will be unavoidable under the probable enforcement conditions for the specific area under design.

Foundation Drainage (Inflow). Foundation drainage should be barred from sanitary sewer systems by adequate regulations, and as with roof and yard drainage, it should be diverted to a storm sewer system. Again, complete enforcement of regulations seldom occurs and allowances must be made for illegal connections. Expected quantities of flow from foundation drain connections may vary considerably. Nevertheless, they must be evaluated for each system.

b. Infiltration

Sanitary sewers must be designed to carry unavoidable amounts of groundwater infiltration or seepage in addition to the peak sanitary flows and unexcludable quantities of stormwater (inflow).

Groundwater gains entrance to sewers as infiltration through pipe joints, broken pipe, cracks or openings in manholes, and similar faults. Defective service connections also can contribute appreciable quantities of infiltration.

Prior to the use of modern compression-type gasket pipe joints, the bulk of infiltration in structurally sound pipe entered at inefficient pipe joints. Many sanitary sewers have been built with either cement-mortar, hot-poured or cold-installed bituminous joints. None of these pipe jointing materials is satisfactory because of the initial difficulty in making a tight pipe joint and the normal deterioration with time. Fortunately modern pipe jointing practice and the use of compression-type gasket pipe joints make it possible to reduce leakage significantly from this source. Most leakage into new sanitary sewer systems now can be traced to defects in pipe foundations or pipe strengths, faulty installation practices or service connections. (A detailed discussion of pipe joints and pipe jointing materials is found in Chapter 8.)

Poorly laid service connections may be extremely important sources of excessive infiltration since these lines often have a total length greater than the collecting sanitary sewers. Service connections have been found to contribute as much as 90% of the total infiltration into the sanitary system. Lack of quality inspection and workmanship in the installation of private service connections has resulted in some cities requiring pressure tests on the completed connection (10). The need for suitable public control of those connections in every community, including detailed specifications, proper construction practices and inspections, cannot be overemphasized.

Existing sanitary sewerage systems are frequently susceptible to significant infiltration rates. Infiltration rates as high as 140 m^3/day/km (60,000 gpd/mile) of sanitary sewer have been recorded for systems installed below groundwater, with rates up to and exceeding 2,350 m^3/day/km (1 mgd/mile) in isolated segments. (Infiltration and exfiltration tests and allowances for new installations are discussed in Chapter 6.)

As with all other sources of unwanted wastewater, infiltration must be kept to a minimum if the cost of pumping and treating wastewater is to be minimized (10). Excessive amounts of infiltration also can result in the need for increased sewer pipe sizes or for additional sewers to supplement existing lines.

The design of extensions to existing systems should consider past practices and trends in infiltration, with due allowances made where necessary. By far the majority of stipulated allowances fall within the ranges shown in Table 3-7 (11). However, most regulatory agencies allow rates in the range of 10 to 40 L/day/mm diam/km (100 to 400 gpd/in. dia/mile). Comparison of the data in Tables 3-7 and 3-8, however, indicates that specified infiltration allowances have not been reduced significantly in the 10-yr interval between the reports.

Table 3-7 Infiltration Specification Allowances

Pipe Diam (In.) (1)	Infiltration Permitted	
	(gpd/mile) (2)	(gpd/in. diam/mile) (3)
8	3,500 to 5,000	450 to 625
12	4,500 to 6,000	375 to 500
24	10,000 to 12,000	420 to 500

Note: In. × 2.54 = cm; gpd/in. diam/mile × 0.000925 = m³/day/cm diam/km.

Table 3-8 Variation of Infiltration Allowances among Cities

Number of Cities Reporting (1)	Allowance (gpd/in. diam/mile) (2)
4	1,500
4	1,000
1	800
2	700
1	600
63	500
11	450 to 300
16	250 to 150
21	100
5	50

Note: Gpd/in. diam/mile × 0.000925 = m³/day/cm diam/km.

The selection of a capacity allowance to provide for infiltration should be based on the physical characteristics of the tributary area, the type of sewer pipe and pipe joint to be used, and sewer pipes in the existing contributary sanitary sewers. For small to medium-sized sanitary sewers 600 mm (24 in.) in diameter and smaller, it is common to allow 71 m³/day/km (30,000 gpd/mile) for the total length of main sewers, laterals, and house connections, without regard to sanitary sewer size. Others make an allowance of from 24 to 95 m³/day/km (10,000 to 40,000 gpd/mile), depending on sanitary sewer size and job conditions. Regulatory agencies in most states have maximum allowances. In general, the design infiltration allowance is added to the peak rate of flow of wastewater and other components to determine the actual design peak rate of flow for the sanitary sewer.

A survey of municipal infiltration allowances (12) is summarized in Table 3-9.

Table 3-9 Infiltration Design Allowances for Several Cities

City (1)	Allowance (gpd/acre) (2)	Remarks (3)
Seattle, Wash.	1,100	
Bay City, Tex.	1,000	
Lorain, Oh.	1,000	
Marion, Oh.	750	Calculations based on a proposed density, with a 100 gpd/cap average flow and peak of 400 gpd/cap also often used.
Ottumwa, Ia.	600	Infiltration and exfiltration shall not exceed 200 gal per inch of pipe diameter per mile of pipe per 24 hr period.
West Springfield, Mass.	2,000	
Alma, Mich.	140	

Note: Gpd/acre × 0.00935 = m³/day/ha; gal × 3.785 = L; in. × 2.54 = cm; miles × 1.61 = km

It is important to note that design allowances for infiltration normally are greater than infiltration-exfiltration test allowances. Infiltration-exfiltration tests are performed when the sanitary sewer is constructed. The design allowance is based normally on the anticipated condition of the sanitary sewer when it is nearing the end of its useful life.

12. Minimum and Peak Flows of Sanitary Wastewater

The flow of wastewater (exclusive of groundwater infiltration and unavoidable inflow) will vary continuously throughout any one day, with extreme low flows usually occurring between 2 and 6 a.m. and peak flows occurring during the daylight hours. The infiltration/inflow component, on the other hand, remains reasonably constant throughout any one day except during and immediately following periods of rainfall.

The ratio of the peak flow of the sanitary component to the average for the day will range from less than 130% for some large sanitary sewers to more than 200% for smaller sanitary lateral sewers. Moreover, the ratio of the maximum daily flow at the end of the design period to the minimum daily flow at the beginning of the period may range from less than two to more than five, depending largely on the rate of growth of the area served by the sanitary sewer (13, 17). Hence, the range of flows for which a sanitary sewer must be designed, that is, peak flow to extreme minimum, will vary from less than 3:1 for large sanitary sewers serving stable populations to more than 20:1 for small sanitary sewers serving growing populations where domestic wastewater is the major component of the total flow. The ratios may be much greater where infiltration/inflow are the governing factors.

The records of existing wastewater or water systems are rarely complete enough to permit estimates of the sanitary sewage component of the minimum flow or peak flow. On a broader basis, Figs. 3-4, 3-5, 3-6 and 3-7 are examples of the variations in peak and minimum rates of flow for situations in which dry-weather wastewater flows are expected to govern. Fig. 3-4 shows the ratios of peak and minimum flows to average daily wastewater flow recommended for use in design by various authorities. Fig. 3-5, based on dry-weather maximums, is the modification of a chart originally prepared for the design of sanitary sewers for a group of 18 cities and towns in the Merrimack River Valley, Massachusetts. The ratios given are approximately correct for a number of other municipalities in the same general area. Fig. 3-6 was developed by the Bureau of Engineering, City of Los Angeles, California, and has been in use since 1962. Fig. 3-7 shows peak residential wastewater flow for the city of Toronto, Canada.

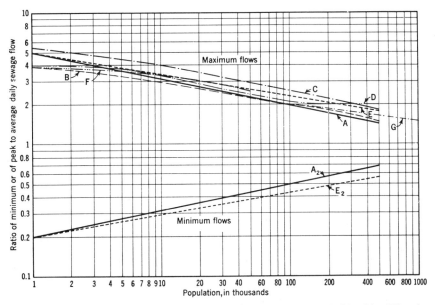

* Curve A source: Babbitt, H. E., "Sewerage and Sewage Treatment." 7th Ed., John Wiley & Sons, Inc., New York (1953).
Curve A_2 source: Babbitt, H. E., and Baumann, E. R., "Sewerage and Sewage Treatment." 8th Ed., John Wiley & Sons, Inc., New York (1958).
Curve B source: Harman, W. G., "Forecasting Sewage at Toledo under Dry-Weather Conditions." *Eng. News-Rec.* **80**, 1233 (1918).
Curve C source: Youngstown, Ohio, report.
Curve D source: Maryland State Department of Health curve prepared in 1914. In "Handbook of Applied Hydraulics." 2nd Ed., McGraw-Hill Book Co., New York (1952).
Curve E source: Gifft, H. M., "Estimating Variations in Domsetic Sewage Flows." *Waterworks and Sewerage*, **92**, 175 (1945).
Curve F source: "Manual of Military Construction." Corps of Engineers, United States Army, Washington, D.C.
Curve G source: Fair. G. M., and Geyer, J. C., "Water Supply and Waste-Water Disposal." 1st Ed., John Wiley & Sons, Inc., New York (1954).
Curves A_2, B, and G were constructed as follows:

Curve A_2, $\dfrac{5}{P^{0.167}}$

Curve B, $\dfrac{14}{4+\sqrt{P}} + 1$

Curve G, $\dfrac{18+\sqrt{P}}{4+\sqrt{P}}$

in which P equals population in thousands.

Fig. 3-4. Ratio of extreme flows to average daily flow compiled from various sources.

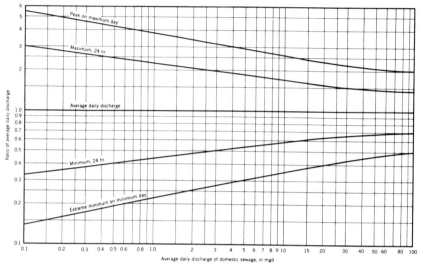

Fig. 3-5. Ratio of extreme flows to average daily flow in New England (mgd × 3.8 = m³/day).

Many state regulatory agenices have established general design parameters of 1.5 m³/day/cap (400 gpd/cap) for laterals and 0.95 m³/day/cap (250 gpd/cap) for trunk sanitary sewers as the minimum acceptable design flow rates (average daily flow per capita) where no actual measurements or other pertinent data are available. These minimum values assume the presence of a normal quantity of infiltration but make no allowance for flows from foundation drains, roofs, yard drains, or unpolluted cooling water. Additional design quantities should be added where conditions favoring excessive infiltration or inflow are present. Also, provision must be made for industrial wastes which are to be transported by the sanitary sewers.

a. Fixture-Unit Method of Design

Estimate of peak sewage flows for facilities such as hospitals, hotels, schools, apartment buildings and office buildings may be made by the "fixture-unit" method (14). Flows from these facilities approach peak rates during the daylight hours. If the velocities in sanitary sewers designed for these flows are adequate for self-cleansing, deposits during the night hours will be resuspended and no nuisance should result. The National Standard Plumbing Code — 1980 (18) defines *fixture-unit flow rate* as "the total discharge flow in gallons per minute of a single fixture divided by 7.5 which provides the flow rate of that particular plumbing fixture as a unit of flow. Fixtures are rated as multiples of this unit of flow." It further defines *fixture-unit* as a "quantity in terms of which the load-producing effects on the plumbing system of different kinds of plumbing fixtures are expressed on some arbitrarily chosen scale." From the first of these it can be seen that a fixture-unit is approximately 0.028 m³/min (1 cfm).

Table 3-10 shows the fixture-unit value for various plumbing fixtures and groups of fixtures. Based on these, the discharge rate for the average single-family house or apartment is about 12 fixture units or three per person for a family of four.

QUANTITY OF WASTEWATER

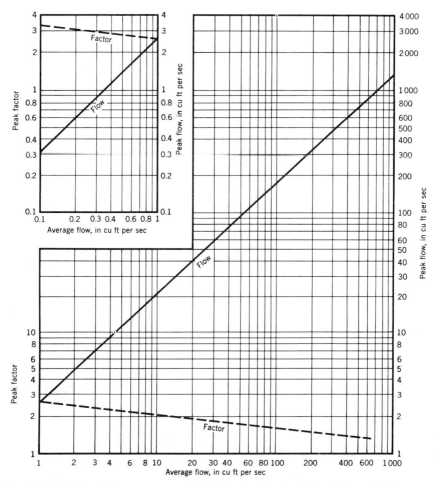

Fig. 3-6. Ratio of peak flow to average daily flow in Los Angeles (cfs × 1.7 = m³/min).

Fig. 3-8 shows the probable peak rates of discharge from systems consisting of various numbers of fixture-units, as taken from probability studies by Hunter (14). The probable peak rate of discharge from a system serving 1,000 persons at three fixture-units per person is, for example, about 28 L/sec (440 gpm). This compares with results obtained from Fig. 3-5 for a 1,000-person system at an average daily discharge of 380 L/cap (100 gal/cap). The peak rate of discharge estimated from Fig. 3-5 is approximately 2,120 m³/day (0.56 mgd), or 25 L/sec (390 gpm).

Although the preceding example exhibits close agreement between a peak flow estimated by the fixture-unit method and that computed on a per capita flow basis, it must be remembered that the fixture-unit method is based on a probability projection. This projection contains a number of assumptions as to the average number of fixture-units and the average water use per capita. It also is based on a distribution of water use or water use habits representing an average or normal population. Large variations in conditions such as the

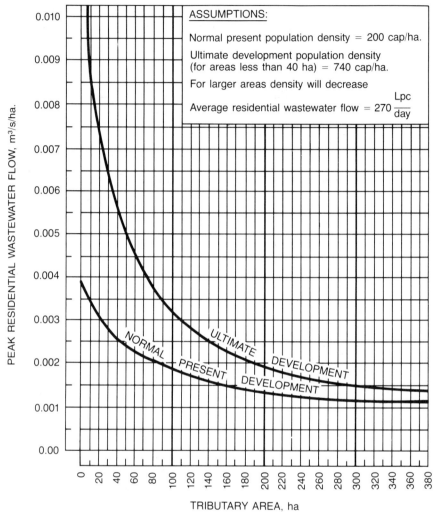

Fig. 3-7. Peak residential wastewater flow for the City of Toronto, Ontario, Canada (m³/sec/ha × 14.5 = cfs/acre; L/cap/day × 0.26 = gal/cap/day; ha × 2.5 = acre).

number of fixture-units per capita and the average use per capita will offset the flow per fixture-unit and cause actual flows to vary from the flows that would be estimated from Fig. 3-8. In the same manner, an extremely homogenous population or a highly regulated population can cause a marked variation between the flows experienced and those predicted.

Maximum water usage in apartment projects in Chicago, Illinois (15), was studied and found to be considerably below that predicted by the fixture-unit method. The results of this study, which covered apartment projects for the elderly, for low income, large family groups, and middle-income families, are shown in Table 3-11.

Table 3-10 Drainage Fixture Unit Values for Various Plumbing Fixtures (18)

Type of Fixture or Group of Fixtures (1)	Drainage Fixture Unit Value (d.f.u.) (2)
Automatic clothes washer (2 in. standpipe)	3
Bathroom group consisting of a water closet, lavatory and bathtub or shower stall	6
Bathtub[a] (with or without overhead shower)	2
Bidet	1
Clinic Sink	6
Combination sink-and-tray with food waste grinder	4
Combination sink-and-tray with one 1½ in. trap	2
Combination sink-and-tray with separate 1½ in. traps	3
Dental unit or cuspidor	1
Dental lavatory	1
Drinking fountain	½
Dishwasher, domestic	2
Floor drains with 2 in. waste	3
Kitchen sink, domestic, with one 1½ in. trap	2
Kitchen sink, domestic, with food waste grinder	2
Kitchen sink, domestic, with food waste grinder and dishwasher 1½ in. trap	3
Kitchen sink, domestic, with dishwasher 1½ in. trap	3
Lavatory with 1¼ in. waste	1
Laundry tray (1 or 2 compartments)	2
Shower stall, domestic	2
Showers (group) per head[b]	2
Sinks:	
Surgeon's	3
Flushing rim (with valve)	6
Service (trap standard)	3
Service (P trap)	2
Pot, scullery, etc.[b]	4
Urinal, pedestal, syphon jet blowout	6
Urinal, wall lip	4
Urinal, stall, washout	4
Urinal trough (each 6-ft. section)	2
Wash sink (circular or multiple) each set of faucets	2
Water closet, private	4
Water closet, public	6
Fixtures not listed above:	
Trap size 1¼ in. or less	1
Trap size 1½ in.	2
Trap size 2 in.	3
Trap size 2½ in.	4
Trap size 3 in.	5
Trap size 4 in.	6

[a] A shower head over a bathtub does not increase the fixture unit value.

[b] See Section 11.4.2 of Ref. 18 for method of computing equivalent fixture unit values for devices or equipment which discharge continuous or semi-continuous flows into sanitary drainage systems.

Note: 1 in. = 2.54 cm; 1 ft = 0.305 m

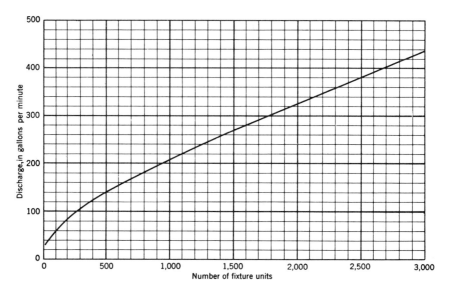

Fig. 3-8. Relation of peak discharge to fixture-units (gpm × 0.0631 = L/sec).

D. RECAPITULATION

The quantity of wastewater which must be transported is based on full consideration of the following:

(1) The design period during which the predicted maximum flow will not be exceeded.

(2) Domestic wastewater flow projections should be based on future population and future per capita water consumption projections. However, unless a more precise parameter than water consumption is available, careful analysis should be made of population distributions and the relationship of maximum and minimum to average per capita wastewater flows. The fixture-unit method of estimating peak rates should be employed for small populations, giving due care to estimating the probable number of fixture-units and water use per capita. When large areas are to be considered, the peak rate of flow per capita or per acre sometimes is decreased as areas and population increase.

(3) Commercial area contributions are sometimes assumed to be adequately provided for in the peak allowance for per capita wastewater flows in small communities. A per acre allowance for comparable commercial areas based on records is the more reasonable approach for large communities.

(4) Industrial wastewater flows should include the estimated employee contribution estimated or gauged allowances per acre for industry as a whole, and estimated or actual flow rates from plants with process wastewaters which may be permitted to enter the sanitary sewer.

(5) Institutional wastewaters are usually domestic in nature, although some industrial wastewaters may be generated by manufacturing operations at prisons, rehabilitation centers, etc.

(6) Air conditioning and industrial cooling waters, if permitted to enter sanitary sewers, may amount to 5.7 to 7.6 L/min (1.5 to 2.0 gpm) per ton of nonwater-conserving cooling units. Unpolluted cooling waters should be kept out of sanitary sewers.

QUANTITY OF WASTEWATER 45

Table 3-11 Maximum Water Demands in Selected Chicago Apartments*

No. Apts. (1)	Population		Fixture Units			Maximum Demand (gpm)							
						Predicted by Hunter Curve (14)				Revised† or Observed Demands			Per Fixture Unit (13)
	Total (2)	Per Apt (3)	Total (4)	Per Apt (5)	Per Cap (6)	Total (7)	Per Apt (8)	Per Cap (9)	Total (10)	Per Apt (11)	Per Cap (12)		
a. Apartments for Elderly													
116	191	1.65	1,400	12.1	7.3	240	2.07	1.26	125	1.08	0.66		0.089
181	299	1.65	2,000	11.1	6.7	320	1.77	1.07	200	1.10	0.67		0.100
252	416	1.65	2,850	11.3	6.8	410	1.63	0.99	280	1.11	0.66		0.098
129	213	1.65	1,200	9.3	5.6	260	2.02	1.22	160	1.24	0.75		0.133
198	327	1.65	2,175	11.0	6.6	350	1.77	1.07	220	1.11	0.67		0.101
151	249	1.65	1,500	9.0	6.0	290	1.92	1.16	175	1.16	0.71		0.117
200	330	1.65	2,200	10.0	6.7	350	1.75	1.06	185	0.93	0.56		0.084
482	795	1.65	7,200	14.9	9.1	625	1.30	0.79	275	0.57	0.35		0.038
157	259	1.65	1,550	9.9	6.0	300	1.91	1.16	190	1.21	0.73		0.123
151	249	1.65	1,500	9.9	6.0	290	1.92	1.16	175	1.16	0.71		0.117
b. Apartments for Low-Income Families													
480	3,312	6.9	5,280	11.0	1.6	640	1.32	0.194	282	0.59	0.09		0.054
140	728	5.2	1,540	11.0	2.1	275	1.96	0.378	151	1.08	0.21		0.098
474	2,940	6.2	5,214	11.0	1.8	638	1.35	0.217	311	0.66	0.11		0.060
c. Apartments for Middle-Income Families													
318	670	2.1	3,600	11.3	5.4	470	1.48	0.70	307	0.97	0.46		0.085

* Chicago Housing Authority. † Data predicted from test data. Note: Gpm × 0.0631 = L/sec.

(7) Infiltration may occur through defective pipe, pipe joints, and structures. The probable amount should be evaluated carefully. Design allowances should be larger than those stipulated in construction specifications for which acceptance tests are made very soon after construction. Under-evaluation of infiltration is one reason why some sanitary sewers have become overloaded.

(8) Inflow may result from foundation, basement, roof, or areaway drains, or storm runoff entering through manhole covers. Foundation, roof, and areaway drain connections to sanitary sewers should be prohibited. Proper construction and yard grading practices should be mandatory. Nevertheless, there may be times when strict prohibition may not be feasible or even practicable. In any event, some storm and surface water will gain entrance to sanitary sewers, and judgement allowance for the inflow, therefore, must be made.

(9) The relative emphasis given to each of the foregoing factors varies among engineers. Some have developed single values of peak design flow rates for the various classifications of tributary area, thereby integrating all contributory items. It is recommended, however, that maximum and minimum peak flows used for design purposes be developed step by step, giving appropriate consideration to each factor which may influence design.

E. REFERENCES

1. Weller, L. W., and Nelson, M. K., "A Study of Stormwater Infiltration into Sanitary Sewers," *Jour. Water Poll. Control Fed.*, Vol. 35, pg. 762, 1963
2. Weller, L. W., and Nelson, M. K., "Diversion and Treatment of Extraneous Flows in Sanitary Sewers," *Jour. Water Poll. Control Fed.*, Vol. 37, pg. 343, 1965
3. McLean, J. E., "More Accurate Population Estimate by Means of Logistic Curves." *Civil Eng.* Vol. 22, pg. 35, 1952
4. McJunkin, F. E., "Population Forecasting by Sanitary Engineers," *Jour. San. Eng. Div.*, Proc. Amer. Soc. Civil Engr., Vol. 90, SA4, pg. 31, 1964
5. Stanley, W. E., and Kaufman, W. J., "Sewer Capacity Design Practice," *Jour. Boston Soc. Civil Engr.*, pg. 317, 1953
6. Johnson, R. E. L., "Development of Sanitary Sewer Design Criteria for the City of Houston, Texas," *Jour. Water Poll. Control Fed.*, Vol. 37, pg. 1597, 1965
7. Babbit, H. E., *Sewerage and Sewage Treatment*, John Wiley & Sons, Inc., New York, N.Y., 7th Ed., 1953
8. *Federal Register*, Vol. 43, No. 188, 40 CFR 35,900 Appendix A, Sept. 27, 1978.
9. Rawn, A. M., "What Cost Leaking Manhole?" *Waterworks and Sewerage*, Vol, 84, 12, pg. 459, 1937.
10. Horne, R. W., "Control of Infiltration and Storm Flow in the Operation of Sewerage Systems," *Sew. Works Jour.*, Vol. 17, pg. 209, 1945
11. Velzy, C. R. and Sprague, J. M., "Infiltration Specifications and Tests," *Sewage and Industrial Wastes*, Vol. 27, pg. 245, 1955
12. Anon, "Municipal Requirements of Sewer Infiltration," *Pub. Works*, Vol. 99, 6, pg. 158, 1965
13. Geyer, J. C., and Lentz, J. L., "An Evaluation of the Problems of Sanitary Sewer System Design", *Jour. Water Poll. Control Fed.*, Vol. 38, pg. 1138, 1966
14. Hunter, R. B., "Methods of Estimating Loads on Plumbing Systems," Rept. BMS 65, National Bureau of Standards, Washington, D.C., 1940
15. Braxton, J. S., "Water Pressure Boosting Systems, Evaluation of Water Usage and Noises," *Cons. Engr.*, Vol. XXIV, V. pg. 112, 1965
16. *Recommended Standards for Sewage Works*, Great Lakes-Upper Mississippi River Board of State Sanitary Engineers, Health Education Service, Inc., Albany, N.Y. 1978 Revised Edition
17. Young, J. C., Cleasby, J. L., Baumann, E. R., "Flow and Load Variations in Treatment Plant Design," *Journal of the Environmental Engineering Division*, ASCE, Vol. 104, pg. 289, 1978.
18. *National Standard Plumbing Code*-1980, National Assoc. of Plumbing, Heating, Cooling Contractors, Washington D.C.

CHAPTER 4

SULFIDE GENERATION, CORROSION, AND CORROSION PROTECTION IN SANITARY SEWERS

A. INTRODUCTION

A sanitary sewer is considered a potentially corrosive environment where hydrogen sulfide (H_2S) may be generated. H_2S may cause various problems including odor, hazard to maintenance crews, and corrosion of some sanitary sewer pipe materials. In the design of sanitary sewer systems, considerations must be given to sulfide generation and to anticipated exposure to various other corrosive agents.

When considering internal corrosion of sanitary sewer pipe, the design engineer should address the importance of corrosion control through proper hydraulic design, effective maintenance, and reasonable control of wastes entering the sanitary sewer. In general, the design of sanitary sewers seeks to preclude septic conditions and to provide a relatively hydrogen sulfide-free environment. Proper wastewater discharge regulations and civil enforcement must be encouraged to prevent the discharge of corrosive waste chemicals and high-temperature wastewater into sanitary sewer systems; such corrosive agents and abnormal wastewater discharges exceed the design limits of typical sanitary sewers and may either damage sewer materials or obstruct the proper performance of wastewater treatment facilities.

B. SULFIDE GENERATION

Retention of wastewater in anaerobic conditions may result in sulfide generation, a reaction due principally to the bacterial reduction of sulfate contained in the sanitary wastewater. Sanitary sewer force mains and gravity flow lines flowing completely filled provide conditions particularly favorable for sulfide generation. Sulfide generation also takes place in partially full sanitary sewers which convey wastewater at low velocities; in such ambient conditions very little oxygen may be transferred to the wastewater from the atmosphere.

Generation of sulfide occurs within the slime layer found on the interior wall of the sanitary sewer pipe. If substantial sulfide generation is to be prevented in the sanitary sewer, the oxygen supply must be sufficient to maintain a reasonable concentration of oxygen (generally, several tenths of a milligram per liter — ppm) in the wastewater, thereby promoting an aerobic condition in the surface of the slime layer.

Fig. 4-1 illustrates sulfide generation processes in a sanitary sewer when insufficient dissolved oxygen is present in the wastewater. When dissolved oxygen in the wastewater drops to an excessively low level, insufficient oxygen enters the slime surface zone to oxidize all of the sulfide generated in the slime anaerobic zone. H_2S generation then becomes a significant problem.

Sulfide in municipal wastewaters may be present in part as insoluble

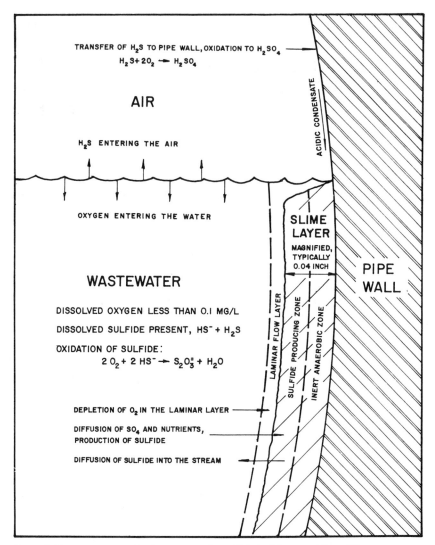

Fig. 4-1. Processes occurring in sewer under sulfide buildup conditions (6).

sulfides of various metals; however, concentration of metal sulfide is generally low, usually a few tenths of a milligram per liter. The major portion of sulfides generated in sanitary sewers is normally retained in solution as a mixture of hydrogen sulfide (H_2S) and ionic HS^-; the mixture is called dissolved sulfide. When the pH of a sulfide-containing wastewater is 7.0, approximately 50% of the dissolved sulfides present will be H_2S; the remainder will be ionic HS^-.

Table 4-1 shows the proportions of H_2S at various pH levels.

The ionization equilibrium of H_2S is shown thus:

$$H_2S \;=\; HS^- \;+\; H^+ \tag{4.1}$$

Table 4-1. Proportions of Dissolved Sulfide Present as H_2S, Assuming pK′ = 7.0

pH (1)	Percent H_2S (2)
5.0	99
6.0	91
6.5	76
7.0	50
7.1	44
7.2	39
7.4	28
7.5	24
7.6	20
7.7	17
7.8	14
7.9	11
8.0	9

The relative proportions of H_2S and HS^- dissolved in wastewater are predicted by the following equation:

$$\frac{[HS^-]}{[H_2S]} = pH - pK' \qquad (4.2)$$

where $[HS^-]$ = molar concentration of the hydrosulfide ion; $[H_2S]$ = molar concentration of hydrogen sulfide; pH = negative logarithm of the activity of the hydrogen ion; and pK′ = negative logarithm of the practical ionization constant in water (average value of 7.0 is adequate for design estimates).

C. FORECASTING SULFIDE CONDITIONS

1. Major Design Considerations — Sulfide Generation

Major design considerations related to the generation of sulfides in sanitary sewer systems are:

(a) High wastewater temperatures are conducive to sulfide generation in sanitary sewers. Substantial sulfide generation is not common in sanitary sewers located in cold climate regions, but it may occur anywhere, and sulfide-free sanitary sewers may be found in hot climate regions.

(b) The reduction of infiltration of groundwater into sewers, without a corresponding modification of other design practices, has sometimes caused the development of sulfide conditions where none previously existed. This is due to reduction of dilution, the resulting increase in BOD and the increase in time-of-travel of the higher-strength wastewater.

(c) Hydraulic flow conditions which relate to slope of the sanitary sewer pipe, the wastewater flow velocity, and the volume of conveyed wastewater will substantially affect sulfide generation. Generally, at very slow flow veloci-

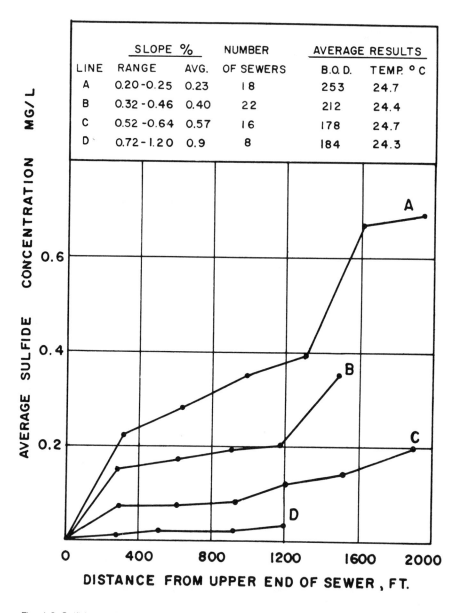

Fig. 4-2. Sulfide occurrence in small sewers in residential areas (6) (6 and 8-in. nominal diameters) (in. × 25.4 = mm; ft × 0.3 = m).

ties when the hydraulic slope is inadequate for the conveyed volume of wastewater, sulfide generation is probable. Sulfide occurrence in typical small sanitary sewers is strongly related to hydraulic slope (see Fig. 4-2).

(d) A permanent deposit of waste solids in the sanitary sewer line will have little effect on sulfide generation, because sulfide generation generally takes place in the surface material, generally less than 1 mm (0.04 in.) deep, of the waste deposit. However, if waste solids are intermittently agitated and

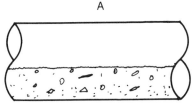

A

Velocity 2 fps, Efficient solids transport. No sulfide buildup in small flows, up to 2 cfs. Sulfide build up often observed in larger flows but only at very slow rate.

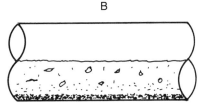

B

Velocity 1.4 to 2.0 fps. Inorganic grit accumulating in the bottom. More sulfide buildup as the velocity diminishes.

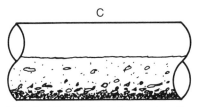

C

Velocity 1.0 to 1.4 fps. Inorganic grit in the bottom, organic solids slowly moving along the bottom. Strongly enhanced sulfide buildup; severe problems expected.

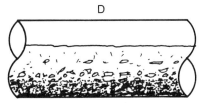

D

Velocity below 1.0 fps. Much organic and inorganic solid matter accumulating, overlain with slow-moving organic solids. Sulfide problems worse than in C.

Fig. 4-3. Solids accumulations at various flow velocities (6) (fps × 0.3 = m/sec; cfs × 0.03 = m³/sec).

moved along the pipe invert, sulfide generation is increased. Fig. 4-3 shows that wastewater flow velocity has significant effect on the transport of solids and the generation of sulfide in a sanitary sewer.

(e) In sanitary sewers with diameters larger than 600 mm (2 ft), sulfide buildup at a slow rate may occur even with velocities above 1 m/sec (3.3 ft/sec) and sometimes up to 2 m/sec (6.6 ft/sec) in very large trunk lines. However, a decline in sulfide content of wastewater in such sanitary sewers may be noted downstream from a point of turbulence (e.g., junction, hydraulic jump, etc.) due to transfer of oxygen and escape of H_2S to the atmosphere.

2. Sulfide Forecast Criteria

a. Flow-Slope Relationship

Figs. 4-4 and 4-5 may be used as a qualitative guide in evaluating sulfide buildup potential; however, they must be qualified as being related to wastewater streams of specified characteristics. In the sanitary sewer, the temperature of the wastewater follows a seasonal cycle, while BOD and quantity of flow follow a diurnal cycle. It is useful to define a climatic condition as the combination of the average temperature for the warmest three months of the year and the average 6-hr high-flow BOD for the day. Where diurnal BOD curves have not been made, it may be assumed that the BOD for the 6-hr high-flow period is 1.25 times the BOD of a flow-proportioned 24-hr

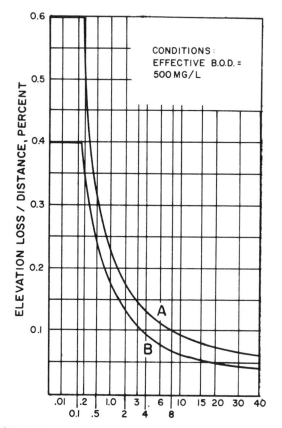

Fig. 4-4. Flow-slope relationships as guides to sulfide forecasting, (6) Effective BOD 500 mg/L (cfs × 0.03 = m³/sec).

composite. The climatic EBOD (effective biochemical oxygen demand) is defined by the equation:

$$(EBOD)_c = (BOD)_c \times 1.07^{(T_c - 20)} \tag{4.3}$$

where $(EBOD)_c$ = climatic EBOD, in milligrams per liter; $(BOD)_c$ = climatic BOD in milligrams per liter; T_c = climatic temperature, in degrees Celsius; and 1.07 = empirical coefficient which may vary with wastewaters having uncommon characteristics.

Figs. 4-4 and 4-5 are intended to provide crude guidelines for anticipating the likelihood of sulfide buildup. If the design of the sanitary sewer falls in the field above Curve A of those figures, it is unlikely that there will be any serious sulfide buildup. If the design falls below Curve B, the development of sulfide is likely. The use of Figs. 4-4 and 4-5 is restricted to sanitary sewers with flow depths not exceeding two-thirds of the pipe inside diameter.

Although the relationship between EBOD and required slope is complex, approximately similar sulfide conditions will result if the effective slopes are increased or decreased in proportion to the square root of EBOD. For example,

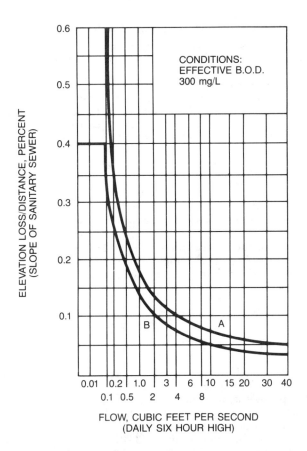

Fig. 4-5. Flow-slope relationships as guides to sulfide forecasting, (6) Effective BOD 300 mg/L (cfs × 0.03 = m³/sec).

for the Curve A condition in Fig. 4-4, a flow of 2 cfs (57 L/sec) requires an effective slope of approximately 0.175%. If the EBOD were 600 mg/L (600 ppm) instead of 500 mg/L (500 ppm), the effective slope for the Curve A condition at 2 cfs (57 L/sec) would be:

$$0.175\% \times \frac{\sqrt{600}}{\sqrt{500}} = 0.19\% = \text{effective slope} \qquad (4.3.a)$$

b. Potential for Sulfide Buildup

Another indicator of the likelihood of sulfide buildup in relatively small gravity sewers (not over 600mm or 24 in.-diameter) is afforded by the "Z" formula (8):

$$Z = \frac{\text{EBOD}}{S^{0.50} \, Q^{0.33}} \times \frac{P}{b} \qquad (4.4)$$

Table 4-2. Sulfide Generation Based on Z Values

Z Values (1)	Sulfide Condition (2)
Z < 5,000	Sulfide rarely generated
5,000 ≤ Z ≤ 10,000	Marginal condition for sulfide generation
Z > 10,000	Sulfide generation common

where Z = defined function; S = hydraulic slope; Q = discharge volume, in cubic feet per second; P = wetted perimeter, in feet; b = surface width, in feet.
Sulfide generation probability is approximated in Table 4-2.

c. Quantitative Sulfide Forecasting — Filled Pipe Condition

In completely filled sewer pipes (force mains) an estimate of sulfide buildup can be obtained using the following equation:

$$\frac{d[S]}{dt} = \frac{3.28 \, M \, [EBOD]}{0.25D} \left(\frac{1 + 0.12 D}{0.12 D} \right) \tag{4.5}$$

in which $d[S]/dt$ is the increase of sulfide concentration in milligrams per liter per hour; M is a coefficient generally taken to be 0.3 mm/hr (0.001 ft/hr) and D is the sewer pipe internal diameter in millimeters (feet). Use of this equation is generally encouraged since force mains are the primary source of serious sulfide problems encountered in sewers.

The use of M equal to 0.3 mm/hr (0.001 ft/hr) is intentionally conservative. Actual buildup will be less than predicted in about 80% of the cases principally because of dissolved oxygen in the wastewater before it enters the main. The use of M equal to 0.3 mm/hr (0.001 ft/hr) is advised unless there is good reason to believe that there will be a substantial delay in the onset of sulfide generation, or if sulfide buildup studies indicate another value is appropriate.

d. Quantitative Sulfide Forecasting — Partially Filled Pipe Conditions

In 1977, Pomeroy and Parkhurst presented a quantitative method for sulfide prediction in gravity (partly filled) sanitary sewers based on continuing research on the wastewater collection systems of Los Angeles County and other data (8). The method is applicable only when conditions are favorable to sulfide buildup. Misleading results may be obtained under other conditions. Sulfide buildup will not occur when the wastewater contains significant dissolved oxygen or when sufficient nutrients are absent as indicated by a low BOD. Therefore, the following equation should not be used for systems where dissolved oxygen exceeds 0.5 mg/L or for sanitary sewers with an effective slope greater than 0.6%:

$$\frac{d[S]}{dt} = M' \, [EBOD] \, r^{-1} - N(sv)^{0.375} \, [S] \, d_m^{-1} \tag{4.6}$$

where $d[S]/dt$ = rate of change of total sulfide concentration, in milligrams per liter-hour; M' = effective sulfide flux coefficient, in meters per hour; r = hydraulic radius of the stream, in meters; N = empirical coefficient; [S] = total sulfide concentration, in milligrams per liter; s = energy gradient of the stream; v = velocity, in meters per second; and d_m = mean hydraulic depth, in feet.

The coefficients M' and N may be established to suit the conditions of the system under consideration. M', however, is usually about 0.4×10^{-3} m/hr (1.3×10^{-3} ft/hr) when dissolved oxygen is low (less than 0.5 mg/L) and approaches zero as dissolved oxygen concentrations increase. The sulfide loss coefficient, N, is a parameter for sulfide losses due to oxidation and escape of H_2S to the atmosphere. Pomeroy has suggested values of M' of 0.32×10^{-3} m/hr (1.0×10^{-3} ft/hr) and for N, either a "conservative" value of 0.64 or a "less conservative" value of 0.96. Analysis of an in-line study led to the view that the less conservative values for N were adequate to predict maximum sulfide levels where major sanitary sewers of several miles (kilometers) in length were involved.

The negative term of Eq. 4.6 is proportional to the sulfide concentration. The sulfide concentration approaches a limiting concentration, $[S]_{lim}$. When Eq. 4.6 equals 0, and taking N equal to 0.96, the equation may be rearranged for the sulfide limiting condition:

$$[S]_{lim} = \frac{0.33 \times 10^{-3} [EBOD]}{(sv)^{0.375}} \times \frac{d_m}{r} \tag{4.7}$$

To calculate the sulfide generation in a long reach with uniform slope and flow, the following integrated form of Eq. 4.7 can be applied:

$$[S]_2 = [S]_{lim} - \left\{ \frac{[S]_{lim} - [S]_1}{\log^{-1}\left[\frac{(sv)^{0.375} \Delta t}{1.15 \, d_m}\right]} \right\} \tag{4.8}$$

where Δt = time of flow from upstream location to downstream location; $[S]_1$ = sulfide concentration at upstream location, in milligrams per liter; and $[S]_2$ = sulfide concentration at downstream location, in milligrams per liter.

D. IN-LINE SULFIDE STUDIES

The sewers selected for in-line studies should be visually inspected. Information including pipe material, extent of any corrosion of the sewer pipe, manhole and manhole steps, sediment and slime depositions, odors and evidence of manhole surcharging should be noted.

Upon selection of the number and location of the sampling points, a monitoring program to determine the change in wastewater properties and composition as it flows through the sanitary sewer may be initiated. Samples must be obtained from the same volume of wastewater as it passes the various sampling locations.

An acceptable period of wastewater sampling should encompass the full range of wastewater temperatures. However, if time and manpower are limited, sulfide buildup tests should be conducted during the period of the maximum 6-hr flows and when wastewater temperatures are at or near the maximum. The number of test runs required depends on the reliability of test data. A minimum of three test runs is recommended. During times of rapidly changing temperatures or BOD the sampling frequency should be increased.

Several parameters must be monitored if the wastewater is to be characterized adequately. These include the BOD_5, pH, temperature, dissolved oxygen, dissolved sulfide concentrations, and flow quantity. The BOD_5 concentration may be measured by laboratory tests on each wastewater sample or on composites. The remaining five parameters must be measured immediately after obtaining the sample.

E. DESIGN CONSIDERATIONS — EFFECTS OF SULFIDES

Major design considerations related to the effects of sulfides in sanitary sewer systems are:

(1) Sulfides present in solution in wastewater conveyed to an activated sludge wastewater treatment plant may impede proper treatment. Increased dissolved sulfide in wastewater reaching a treatment plant increases prechlorine demand. Odor emanating from the influent structure may be a cause of public complaint.

(2) Hydrogen sulfide (H_2S) may result in severe corrosive conditions for unprotected sewer pipes produced from cementitious materials and metals. Such corrosive conditions occur when sulfuric acid (H_2SO_4) is derived through the oxidation of hydrogen sulfide by bacterial action on the exposed sewer pipe wall.

(3) If H_2SO_4 is produced, the inside pipe wall above the wastewater flow line will be the principal area of corrosion. The corrosive effects of sulfuric acid in sanitary sewers will vary according to the type of pipe material used, the concentration of acid present, and the ambient temperature. Concrete pipes, asbestos-cement pipes, and mortar linings on ferrous pipes will experience surface reaction in which the surface material is converted to an expanding, pasty mass which may fall away and expose new surfaces to corrosive attack (see Fig. 4-6). Ferrous pipe materials may experience surface reaction in which a portion of the material is dissolved and a portion is converted to iron sulfides, yielding a hard bulky mass that forms on the exposed surface.

(4) Levels of annual average dissolved sulfide concentration in small sanitary sewers may be 0.1 to 0.2 mg/L without significant probability of severely corrosive conditions, even though substantial corrosion may be seen at points of high turbulence. Sulfide levels from 0.5 mg/L and higher may be tolerated in large sanitary sewers with uniform, laminar flow conditions. At points of high turbulence, however, H_2S is released more rapidly and can result in severe corrosion even when annual average dissolved sulfide does not exceed a few tenths of a milligram per liter.

(5) H_2S gas is extremely toxic. Because the gas is quite common in nature at various low level concentrations, its extremely hazardous nature is frequently ignored. Wastewater containing 2 mg/L of dissolved sulfide and at a pH of 7.0, if brought to equilibrium with air in a closed space, will produce a lethal atmosphere. Deaths have resulted from an H_2S concentration of as low as 0.03% (300 ppm) in the air. Numerous deaths have been caused by H_2S poisoning in sanitary sewer manholes. Hydrogen sulfide is treacherous in that

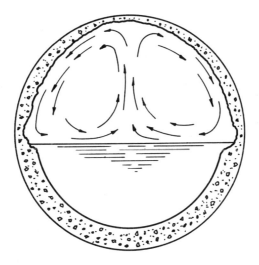

Fig. 4-6. Unequal distribution of corrosion in sanitary sewer. (6)

if a potential victim ignores the first odor of the gas, his ability to smell the gas is quickly lost, thereby eliminating further warning.

(6) H_2S is usually the major component of sanitary sewer system odors. The threshold concentration in water for human detection is between 0.01 and 0.1 mg/L.

(7) Severe corrosion and odor problems can occur in structures where force mains or full-flowing gravity sewers discharge.

Whether a sanitary sewer system is essentially free of sulfides or is the locus of odor and corrosion problems depends to a great extent on the design and operation of the system. It is often impractical or impossible to design a sulfide-free system. In the system design, the engineer should consider control of sulfide generation through system design, corrosion protection, and use of corrosion-resistant pipe materials.

F. SULFIDE CONTROL

Control of sulfide generation through sanitary sewer system design requires consideration of various commonly used control measures.

(1) Wastewater flow velocity should be established at a level which will provide efficient transport of the solids, not just when ultimate design flows are reached but early in the life of the sewer as well.

(2) If dissolved sulfide in the wastewater is likely to be more than 0.2 mg/L, the system design should, where possible, preclude or minimize turbulent flow conditions. Although turbulence can effect a reduction in sulfide generation in the wastewater, release of dissolved H_2S at the point of turbulence will be relatively severe. Where wastewater is fresh, drops and turbulence can be used to maintain a good dissolved oxygen level in the flow and prevent sulfide buildup.

(3) Hydrogen sulfide generation can be controlled effectively through various forms of chemical treatment, but such methods of control are relatively expensive. In the treatment plant, chlorination with either elemental chlorine

or hypochlorite quickly destroys sulfide and odorous organic sulfur compounds. However, chlorination in sanitary collection sewers is generally considered impractical.

Various metal salts, such as ferrous sulfate or other iron salts added to sanitary wastewater, may convert sulfides to insoluble forms. Such treatment is practical for high concentrations of dissolved sulfides; however, it will not accomplish complete elimination of dissolved sulfides.

Hydrogen peroxide (H_2O_2) can be used for sulfide control. Effective use of H_2O_2 can reduce dissolved sulfides to 0.1 mg/L or lower. The peroxide reacts in the wastewater environment by breaking down to oxygen and water, or it reacts directly when sulfide concentrations are high. The presence of oxygen effectively limits sulfide generation.

The dissolving of air or oxygen in the wastewater stream is an effective sulfide control measure. There are two large installations in Sacramento County, Calif., using oxygen in sewers, and this method is extensively used in Europe and in other places.

Addition of sodium hydroxide to a sewer to produce a pH of 12.5 to 13 for 20 minutes will inactivate sulfide producing slimes. Where this method is used, treatments are generally made weekly in the summer, but perhaps only monthly in the winter. It is a procedure that becomes uneconomical in large flows.

(4) Sulfide generation in sanitary sewer force mains can often be prevented by injecting air into the mains. Air injection provides maximum effective control when the force main has a continually rising profile with sufficient slope to promote effective distribution of air bubbles in the wastewater. Pressure mains vulnerable to acid can suffer severe corrosion from air injection if the profile is irregular and an air pocket is formed where H_2S is present. However, an irregular profile does not always preclude air injection. A careful study of an irregular main may show that air injection can be safe and effective.

(5) Limited benefit is sometimes gained by ventilation of sanitary sewers. This method does not significantly reduce sulfide generation; however, the ventilation does remove a portion of free H_2S from the sanitary sewer atmosphere. Ventilation may also dry the walls of structures, and thereby may prevent the conversion of H_2S to H_2SO_4.

G. OTHER CORROSIVE CONDITIONS

Various pipe materials exhibit resistance to corrosive attack from sulfuric acid (H_2SO_4) in sanitary sewers; however, other forms of chemical corrosion must also be considered. Certain concentrated organic solvents can soften the polymeric materials used in plastic pipes and in plastic joints on non-plastic pipes, but damage of this type is very rare.

Ferrous pipe materials immersed in sanitary wastewater generally corrode less than when exposed to potable water because of a diminished dissolved oxygen concentration and an inhibiting effect of the organic matter in the water. A protective lining or coating may reduce tuberculation in the course of the expected life of a sanitary sewer. In a pressure sanitary sewer main, where lines flow full thus preventing the generation of H_2SO_4, there is generally no internal corrosion of the sewer pipe.

The prime causes of external corrosion of buried metal pipe are galvanic action and stray currents.

Galvanic action is the cause of most corrosion in buried iron and steel pipe. The term describes a series of interrelated electrochemical reactions. During the galvanic process, electrons are released which create a direct current of electricity, and corrosion occurs because this galvanic action also causes metal to be removed from the pipe surface.

Stray electric currents also cause external corrosion of iron and steel pipe, but by a process that differs from galvanic action. Stray currents are common in the ground of urban areas. These currents are generated by electric railways, subways, or any electric system grounded to the earth. Such systems include impressed current cathodic protection which is commonly used more and more today to protect a variety of underground utilities and structures. The stray current corrosion process is similar to that involved in the generation of electricity. When a metal pipeline is buried in a field of stray currents, it picks up the currents which then travel along the pipe to a point of lower potential, where the current is discharged. In this case, the discharge point is the anode. The current pick-up area can be rather large but the discharge point is very local, which produces rapid metal loss resulting in pitting and perforation.

The question of possible sulfate attack on cementitious materials has been raised in connection with the use of concrete sanitary sewers. Sulfate attack is a process that occurs only in the strongly alkaline condition that exists in the interior of various types of concrete and only in the presence of high sulfate concentrations. It does not occur in concrete immersed in sea water (normally containing 2,650 mg/L of sulfate) although concrete intermittently wetted with sea water and dried may possibly be affected, and concrete that is not sound may be affected by magnesium ions in sea water. The attack on concrete by sulfuric acid produced from hydrogen sulfide is due to the action of hydrogen ions. In this corrosive condition, concurrent sulfate attack is not possible.

H. CORROSION PROTECTION

With consideration of the corrosive conditions anticipated in a specific system the design engineer may choose to specify corrosion resistant materials or various forms of corrosion protection. Corrosion may be controlled or limited using corrosion protection which generally falls into the categories of linings or coatings, composition of materials and/or thickness of sewer pipe materials. Metallic sewer pipe may also be protected from exterior wall corrosion through the use of suitable concrete encasement, insulating wrappings and/or cathodic protection.

1. Linings and Coatings

To be cost effective, protective linings and coatings must provide effective isolation of the sewer pipe and pipe joints from aggressive agents, thereby permitting the sewer pipe to perform adequately for the design life of the system. United States Environmental Protection Agency (USEPA) cost-effective guidelines require that a service life of 50 yr be used when evaluating funding of interceptor/collector sanitary sewers. The effectiveness of linings and coatings depends not only on the physical and chemical properties of the material used but also on the care taken in manufacture, handling, installation, and maintenance.

Linings for concrete pipe are fixed to the pipe's interior wall by projections embedded in the concrete at the time of pipe manufacture. Linings of sufficient thickness (generally 1.5mm — 0.06 in.) keyed to the interior wall of concrete pipe, if properly handled and installed with fused joints, have provided adequate corrosion protection. Care must be taken to match properly the lining and its application to the pipe manufacturing method and end-use.

Coating materials fixed to the interior wall by adhesion appear, in some cases, to have provided long-term corrosion protection in hydrogen sulfide environments. Experience with specific coatings should be investigated prior to specification. Linings and coatings available for sanitary sewer pipe are as follows:

(a) *Concrete Pipe.* Various linings and coatings are available. These include polyethylene and polyvinyl chloride sheet keyed to the wall when the pipe is cast and coal tar epoxy or similar material applied to the interior surfaces of the pipe. Experience with protective coatings has varied widely.

(b) *Ductile Iron Pipe.* A cement lining with an asphaltic seal coating often is provided on the interior wall of the pipe. In the presence of hydrogen sulfide this lining will be subject to the same corrosive action as cementitious pipe. Various forms of thermoplastic linings fixed by adhesion to the interior wall of the pipe may be provided. An exterior asphaltic coating is commonly provided during manufacture. Where necessary, polyethylene film encasement may be applied during installation around the exterior of the pipe to impede external corrosion.

(c) *Steel Pipe.* Corrugated steel pipe may be provided with various coatings, including asbestos filled materials, fixed by adhesion to the wall.

(d) *ABS Composite Pipe.* Placement of a coating of solvent cement over the exposed ends of the pipe may be made at the time of installation (if the ends were not factory sealed) to impede acid migration into the cementitious filler in the wall annuli.

2. Composition of Materials and/or Thickness of Pipe Material

Where significant sulfide levels are expected, the corrosion of a cementitious sanitary sewer depends on the material of pipe construction as well as on the rate of release of H_2S from the wastewater stream. The total mass emission from the stream is essentially the mass transfer of H_2S to the pipe wall, since only a small amount escapes entirely from the sanitary sewer. Under typical sewer conditions, excluding shallow, high velocity streams or points of high turbulence, and if all the escaping H_2S is oxidized on the pipe wall, the average flux to the wall, ϕ_{sw}, is equal to the flux from the stream multiplied by the ratio of stream surface area to exposed wall area, or surface width divided by exposed perimeter:

$$\phi_{sw} = 0.45 J \, [DS] \left(\frac{b}{P'}\right)(sv)^{0.375} \tag{4.9}$$

where ϕ_{sw} = hydrogen sulfide flux to pipe wall, in grams per square meter-hour; J = factor relating amount of dissolved sulfide as H_2S to wastewater pH from Table 4-1; $[DS]$ = dissolved sulfide concentration in the wastewater, in milligrams per liter; b = surface width of flow stream, in feet; P' = perimeter of pipe exposed to atmosphere, in feet; s = slope of energy line; and v = flow velocity, in feet per second.

Values for ϕ_{sw} can be obtained using Eq. 4.9 or from Fig. 4-7. The following example illustrates the calculations involved in determining the sulfide buildup in a gravity sewer and the sulfide flux to the pipe wall.

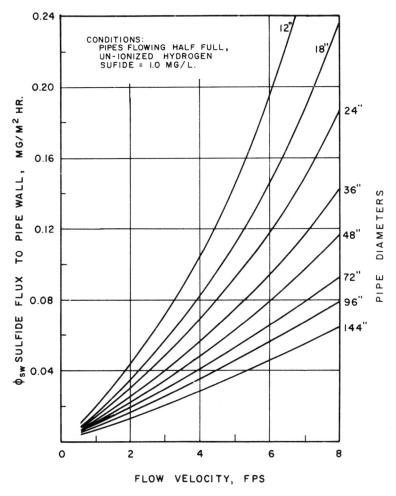

Fig. 4-7. Effect of velocity and pipe size on sulfide flux to pipe wall under specified conditions (m² × 11 = ft²; fps × 0.3 = m/sec; in. × 25.4 = mm; L × 3.8 = gal).

Example 4-1:

A 30-in. (750-mm) diameter pipe sanitary sewer 1,500 ft (460 m) long has been designed to carry 0.70 cfs (0.02 m³/sec) at a relative depth of 0.15 on a 0.1% slope. Mannings "n" is 0.12. The wastewater characteristics are BOD_c, 305mg/L; climatic temperature, T_c, 26°C; pH, 7.2; insoluble sulfide, 0.2 mg/L and total sulfide at the beginning of the reach, S_1, 1.0 mg/L. The sewer hydraulic characteristics are: Surface width of flow, B, 1.79 ft (0.54m); wetted perimeter, P, 1.99 ft (0.60m); exposed perimeter, P', 5.86 ft (1.78m); velocity, v, 1.52 fps (0.46 m/sec); and mean hydraulic depth, d_m, 0.26 ft (0.08m).

Solving Eq. 4.3 for the climatic EBOD:

$[EBOD]_c = [BOD]_c \times 1.07^{(T_c - 20)}$

$= 305 \times 1.07^{(26-20)}$

$= 458$ mg/L

Solving Eq. 4.8 for the limiting total sulfide value gives the following equation:

$$(S)_{lim} = \frac{0.52 \times 10^{-3} [EBOD]}{(sv)^{0.375}} \cdot \frac{P}{b} \quad (4.10)$$

$$= \frac{0.52 \times 10^{-3} \times 458}{(.001 \times 1.52)^{0.375}} \cdot \frac{1.99}{1.79}$$

$$= 3.00 \text{ mg/L}$$

The sewer flow through time is the sanitary sewer length divided by the velocity:

$$\Delta t = \frac{\text{Length}}{v} \quad (4.11)$$

$$= \frac{1500}{1.52}$$

$$= 987 \text{ sec, or } 0.27 \text{ hr}$$

Solving Eq. 4.8 for the total sulfide at the end of the reach:

$$S_2 = (S)_{lim} - \left\{ \frac{(S)_{lim} - S_1}{\log^{-1}\left[\frac{(sv)^{0.375} \Delta t}{1.15 \, d_m}\right]} \right\}$$

$$= 3.00 - \left\{ \frac{3.00 - 1.0}{\log^{-1}\left[\frac{(0.001 \times 1.52)^{0.375} \times 0.27}{1.15 \times 0.26}\right]} \right\}$$

$$= 1.33 \text{ mg/L}$$

The amount of dissolved sulfide is the total sulfide at the end of the reach minus the insoluble sulfide:

$$[DS] = S_2 - \text{(insoluble sulfide)}$$

$$= 1.33 - 0.20$$

$$= 1.13 \text{ mg/L}$$

From Table 4-1, for a pH of 7.2, the J factor is 0.39. Solving Eq. 4.9 for the hydrogen sulfide flux:

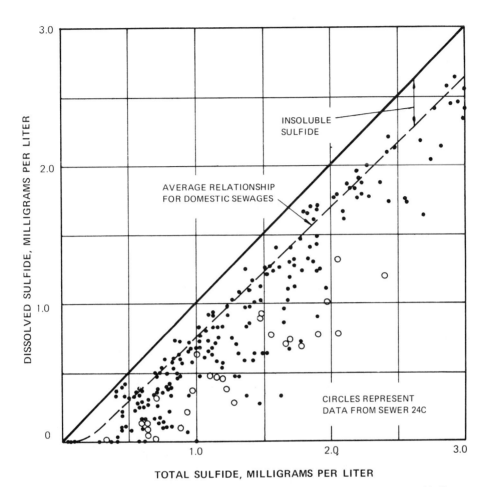

Fig. 4-8. Comparison of total and dissolved sulfide concentrations in trunk sewers, Na₂S added in most cases (6) (L × 3.8 = gal).

$$\phi_{sw} = 0.45 \, J \, [DS] \left(\frac{b}{P'}\right) (sv)^{0.375}$$

$$= 0.45 \times 0.39 \times 1.13 \left(\frac{1.79}{5.86}\right)(0.001 \times 1.52)^{0.375}$$

$$= 0.0053 \text{ g/m}^2 \cdot \text{hr}$$

Fig. 4-7 shows rates of sulfide flux to the pipe wall in pipes flowing half full. The strong upward curvature of the curves for the smaller pipe is due to increased turbulence at high Froude numbers. Fig. 4-8 provides factors to apply to rates from Fig. 4-7 to calculate the flux at other relative flow depths. The velocity input for use in conjunction with Fig. 4-7 must be the velocity that would prevail if the quantity of flow were such as to half-fill the pipe, not the actual velocity for the actual quantity of flow.

(a) *Concrete Pipe.* The rate of corrosion of cementitious material can be estimated from the amount of reactive material in the pipe wall that will consume acid and the rate of acid production. The basic corrosion rate equation is:

$$c = \frac{0.45k\ \phi_{sw}}{A} \tag{4.12}$$

where c = average exposed pipe wall corrosion rate, in inches per year; $k \leq 1.0$, a constant for the amount of sulfide oxidization and acid runoff; and A = concrete alkalinity, as equivalent calcium carbonate, $CaCO_3$.

No actual measurements of k have been made and the choice of a value is a matter of engineering judgment. The value of k would approach unity when the rate of acid formation is very slow, and it may be as low as 0.3 if acid production is rapid and much condensation is present.

The alkalinity of a concrete with 335 kg/m³ (564 lb/cu yd) of cement varies with the aggregate:
- Granitic aggregate, alkalinity from 0.18 to 0.22;
- 50% calcareous aggregate, alkalinity from 0.4 to 0.6;
- 100% calcareous aggregate, alkalinity 1.0 (assumes aggregate at a purity of 80 to 90%).

When the alkalinity of the component materials for a mix are known, the alkalinity of the resultant concrete can be determined. The analysis requires that the proportion of each material used per unit of concrete be expressed as a percentage of the weight of the cured concrete. This percentage is then multiplied by the appropriate material alkalinity to obtain the alkalinity contributed to the mix by that material. For example, anhydrous cement has an alkalinity equal to about 1.18 times its weight. If 335 kg (739 lb) of cement are used to produce a cubic meter (1.3 cu yd) of concrete weighing 2,436 kg/m³ (4,100 lb/cu yd), the cement represents 14% of the concrete, and therefore contributes an alkalinity of 1.18 × 0.14, or 0.17, to this particular concrete mix. The contributing alkalinity of all other materials, including the water used to make the concrete, would be summed to determine the total alkalinity of the concrete mix.

A recent practice in specifying concrete sewer pipe for corrosive environments has combined the practices of specifying either increased alkalinity or sacrificial concrete to result in the product of the alkalinity times the thickness of cover over the reinforcing steel. By defining the allowable corrosion rate as:

$$c = \frac{Z}{L} \tag{4.13}$$

where Z = thickness of allowable concrete loss, in inches (millimeters); and L = service life or design period in years.

Substituting Eq. 4.13 into Eq. 4.12 and rearranging the resulting equation yields:

$$AZ = 0.45k\phi_{sw} L \qquad (4.14)$$

With the resultant AZ in Eq. 4.14 specified for the concrete sewer pipe, the pipe manufacturer has increased flexibility to produce the pipe on a cost effective basis with consideration of manufacturing process, equipment availability and aggregate source. An example illustrating the required corrosion protection, AZ, for a concrete sewer follows.

Example 4-2:

An interceptor sewer has been sized as a 30-in. (760mm) diameter pipe to carry an average flow of 3.5 cfs (99 L/sec) at 0.21 relative depth. A 50-yr design life and a factor of safety of 1.5 have been selected. Total sulfide, T.S., has been determined to be 3.66 mg/L during the climatic period. The ratio of the annual average sulfide level to the peak during the climatic period has been observed to be 0.3. The insoluble sulfide, I.S., is 0.2 mg/L and the annual average pH has been determined to be 7.4. (If field data are not available, the relationship between total and dissolved sulfides may be estimated from Fig. 4-9.)

For the conditions given, the half-full flow velocity is 5.5 fps (1.7 m/sec). The annual average total sulfide is:

$$\text{AVG. T.S.} = 0.3 \times 3.66 = 1.10 \text{ mg/L}$$

The annual average dissolved sulfide, D.S. is:

$$\text{AVG. D.S.} = \text{T.S.} - \text{I.S.} = 1.10 - 0.20 = 0.90 \text{ mg/L}$$

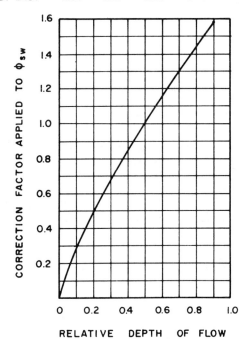

Fig. 4-9. Factor to apply to ϕ_{sw} from Fig. 4-7 to calculate ϕ_{sw} for other than half-pipe depth. (6)

From Fig. 4-7, the uncorrected ϕ_{sw} is 0.09. From Table 4-1, the J factor is 0.28. From Fig. 4-8, the depth correction factor is 0.52.

The correction for dissolved sulfides is 0.90, therefore the corrected ϕ_{sw} is:

Corrected ϕ_{sw} = (uncorrected ϕ_{sw}) (J) (depth correction) (D.S.)
= 0.09 × 0.28 × 0.52 × 0.9 = 0.0115

Assuming a moderate rate of acid production, and setting k equal to 0.7, the required AZ is:

AZ = 0.45k $\phi_{sw}L$
= 0.45 (0.7) (0.0115) (50 × 1.5)
= 0.27

(b) Reinforced Thermosetting Resin (RTR) and Reinforced Plastic Mortar (RPM). In the manufacture of these products, thermoset plastics or thermoplastics may be selected which can substantially increase product resistance to chemical corrosion. It should be noted that wide variation in corrosion resistance can be demonstrated when comparing different plastic materials.

I. REFERENCES

1. Brock, Thomas D., *Biology of Microorganisms*, Prentice-Hall, Englewood Cliffs, New Jersey, 1970.
2. Kienow, Kenneth K. and Kienow, Karl E., "Computer Predicts Pipe Corrosion," *Water and Sewage Works*, September, 1978.
3. Kienow, Kenneth K. and Pomeroy, Richard D., "Corrosion Resistant Design of Sanitary Sewer Pipe," Presented at Session No. 48, ASCE Convention, Chicago, Illinois, October, 1978.
4. Meyers, W.J., "Case Study of Prediction of Sulfide Generation and Corrosion in Sewers." *Journal Water Pollution Control Federation*, 52,11 (Nov. 1980).
5. Parker, C.D., "Mechanics of Corrosion of Cement Sewers by Hydrogen Sulfide," *Sewerage and Industrial Waste*, December, 1951.
6. Pomeroy, Richard D., *Process Design Manual for Sulfide Control in Sanitary Sewers*, Environmental Protection Agency, October, 1974.
7. Pomeroy, Richard D., "Sanitary Sewer Design for Hydrogen Sulfide Control," *Public Works*, 101,10 (Oct. 1970).
8. Pomeroy, Richard D. and Parkhurst, John D., "The Forecasting of Sulfide Buildup Rates in Sewers," *Progress in Water Technology*, Volume 9, Pergamon Press, 1977.
9. *Regulation of Sewer Use, Manual of Practice No. 3*, Water Pollution Control Federation, Washington, D.C. (1979).
10. Romanoff, Melvin, "External Corrosion of Cast-Iron Pipe," *Journal* AWWA, September, 1964.
11. *Standard Methods for Examination of Water and Wastewater*, American Public Health Association, New York, Twelfth Edition, 1965.
12. Sudrabin, L.P., "Protect Pipes From External Corrosion," *American City and County*, May, 1956.
13. *Concrete Pipe Handbook*, American Concrete Pipe Association, 1980.

CHAPTER 5

HYDRAULICS OF SEWERS

A. INTRODUCTION

The purpose of a sanitary sewer is to convey wastewater at various rates of flow. The maximum and minimum flow rates in a single day can vary greatly (see Chapter 3). Furthermore, there is seldom any control over the content of wastewater that must be conveyed to a treatment plant. Wastewaters can contain dissolved solids as well as suspended solids that either settle or float. In general, most of the dissolved solids and floating material are carried along with the flow stream. From the hydraulic point of view, suspended solids that settle along the sanitary sewer pipe invert must be given the greatest consideration because deposition can restrict flow. In unsuitable flow situations, the settleable solids can form deposits that retard the flow and may increase the generation of sulfide. The combination of settled grit and organic material must be removed during the cleaning of the sanitary sewer.

It is clear that the typical sanitary sewer must convey a wide range of flow rates. It should be able to carry the maximum rate of flow intended without backing up to any significant degree. At maximum flow, the velocities and hydraulic forces must not exceed the limits that are imposed by the sewer material. The sewer must also be designed to convey the minimum flow without deposition of suspended solids or with deposition in amounts, at times, and in locations where resuspension will take place with the next increase in flow, before any undesirable effects occur.

The sanitary sewer must accomplish this complex role of conveyance under a severe set of limitations and constraints. For most sanitary sewers the flow should require only the force of gravity and there is a limit as to how much the slope of the sanitary sewer can depart from the surface slope of the ground in which it is located. Thus, the requirements are most often achieved by gravity flow within a restricted range of slopes.

Gravity-flow sanitary sewers are usually designed to flow full or nearly full at peak rates of flow and partly full at lesser flows. Most of the time, the flow surface is exposed to the atmosphere within the sewer, and the sewers function as open channels. At extreme peak flows, some sanitary sewers flow full with wastewater surcharging up into the manholes above the top of the sanitary sewer. The surcharged sanitary sewers function as pressure conduits during those periods. When pressure-conduit flow is permitted as part of a design, it is limited to deep sanitary sewers with no service connections.

In some special cases, sanitary sewers are designed to function as pressure conduits at all times. Pressure sanitary sewers or force mains are regular pressure pipelines that carry wastewater. In addition, there are special sewer applications such as deep tunnels, inverted siphons and submerged outfalls. Pressure conduits are often provided in a sanitary sewer system to overcome some of the restrictions on slope or location that are imposed by gravity flow, such as flow uphill over high topography. However, even this type of sewer must still deal with the great range in flow rates and with deposition in the conduit.

In addition to the overall problems of flow rate and deposition, there are many and varied local problems in a sanitary sewer system. There are frequent changes in pipe slopes, size, and direction. Manholes, special structures to drop flow, junctions, flow controls, structures for measuring flow, and sampling stations for monitoring the quality of the wastewater must be included in the sewer system. All of these appurtenances create local flow situations that must be considered in the hydraulic design to make sure that they do not reduce the sanitary sewer capacity.

The hydraulics necessary to deal with all of the design situations and problems cannot be covered adequately in one chapter. Instead, the approach here is to point out some of the flow problems that occur and to present basic principles and equations that apply to most of them. Some of the problems are discussed in more detail, while others are only mentioned. In all cases, sources of additional discussion and data are indicated.

B. TERMINOLOGY AND SYMBOLS

Insofar as possible the terminology used in this chapter conforms to that given in *Nomenclature for Hydraulics* (1). Further reference to terminology may be found in standard texts (2,3,4,5).

The symbols and units used in this chapter are:

A = cross-section area, in square meters (square feet)
A_f = total area of a closed conduit, in square meters (square feet)
B = dimensionless constant (sediment-scouring characteristic)
b_w = width of water surface, in meters (feet)
C = coefficient, dimensionless; Hazen-Williams resistance coefficient, dimensional
C_f = Chezy coefficient, dimensionless
C_q = discharge coefficient, dimensionless
D = diameter of circular conduit or height of other closed conduit, in meters (feet)
D_g = diameter of sediment grain, in meters (feet)
d = depth of flow above invert, in meters (feet)
d_c = critical depth, in meters (feet)
$\bar{d}$ = depth to centroid of area A, in meters (feet)
d_L = lower stage depth, in meters (feet)
d_m = hydraulic mean depth ($d_m = A/b_w$), in meters (feet)
d_{mc} = hydraulic mean depth at critical flow, in meters (feet)
d_n = normal depth, in meters (feet)
d_U = upper stage depth, in meters (feet)
d_1, d_2 = depths before and after hydraulic jump, in meters (feet)
F = Froude number, dimensionless
$\sum F$ = summation of external forces acting on a fluid body, in newtons (pounds)
f = friction factor, Darcy-Weisbach, dimensionless
f_f = friction factor for conduit flowing full, dimensionless
g = acceleration of gravity, in meters (feet) per second squared
H = total head, in meters (feet)
H_L = head loss, in meters (feet)
ΔH_L = minor head loss, in meters (feet)
H_o = specific energy head, in meters (feet)

h	=	head or piezometric head, in meters (feet)
K	=	minor loss coefficient, dimensionless
k	=	height of wall roughness, in meters (feet)
l	=	length, in meters (feet)
Δl	=	increment of length, in meters (feet)
M	=	symbol for mild slope
M	=	momentum function
n	=	roughness factor (Manning and Kutter), in meters (feet) to one-sixth power
n	=	subscript indicating normal flow
n_f	=	roughness factor (Manning) for conduit flowing full, in meters (feet) to one-sixth power
p	=	pressure, in Pascals (pounds per square foot)
P	=	wetted perimeter, in meters (feet)
Q	=	discharge, in cubic meters (cubic feet) per second
Q_c	=	discharge at critical flow, in cubic meters (cubic feet) per second
Q_n	=	discharge at normal flow, in cubic meters (cubic feet) per second
R	=	Reynolds number, dimensionless
R	=	hydraulic radius (A/P), in meters (feet)
R_f	=	hydraulic radius, full section, in meters (feet)
S	=	slope, dimensionless
S_c	=	critical slope, dimensionless
S_e	=	slope of energy grade line, dimensionless
S_f	=	slope of energy grade line for conduit flowing full, dimensionless
S_0	=	slope of invert or bed, dimensionless
s	=	specific gravity, dimensionless
t	=	time, in seconds
V	=	velocity (mean, Q/A), in meters (feet) per second
V_c	=	critical velocity, in meters (feet) per second
V_f	=	mean velocity of closed conduit flowing full, in meters (feet) per second
V_L	=	supercritical velocity, in meters (feet) per second
V_U	=	subcritical velocity, in meters (feet) per second
V_1, V_2	=	velocity, upper and lower reaches or different conduit sections, in meters (feet) per second
v	=	local velocity, in meters (feet) per second
x	=	distance along sewer, in meters (feet)
y	=	height above invert, in meters (feet)
z	=	height of invert above datum, in meters (feet)
α	=	energy correction factor, dimensionless
β	=	momentum correction factor, dimensionless
γ	=	fluid density (specific weight), in kilograms per cubic meter (pounds per cubic foot)
ν	=	kinematic viscosity, in square meters (square feet) per second
θ	=	slope of sewer invert, dimensionless
τ_0	=	force of friction between fluid and conduit walls, in Pascals (pounds per square foot)

C. HYDRAULIC PRINCIPLES

1. Types of Flow

The discussion in this chapter is concerned with wastewater in sanitary sewers. Hydraulically, however, there is little to distinguish wastewater from stormwater or water. For convenience, the words "water" or "fluid," often used in this chaper to discuss flow, imply wastewater. In the same sense, the word "conduit" is used to indicate any type of sanitary sewer.

The flow of wastewater in a conduit may be either open-channel or pressure flow. In open-channel flow the water surface is exposed to the atmosphere. This type of flow occurs in a sewer as long as any portion of the internal pipe perimeter is not submerged. Pressure flow totally fills a closed conduit.

In general, the movement of wastewater in a sanitary sewer varies with time at any location and with location at any instant of time. If a sanitary sewer has many points of inflow from service connections or smaller branch sanitary sewers, then the rate of flow at any moment varies with distance along the sanitary sewer. Such flow is referred to as spatially varied. If any one of these inflows changes with time, then the flow at any point in the sanitary sewer also changes with time and is referred to as unsteady. In a cross section of the sanitary sewer, the velocity of flow may vary laterally across the section or vertically from the water surface to the invert. The lateral variation is not symmetrical in situations such as a horizontal curve in the sanitary sewer or at the confluence of a large inflow from one side. Because of these variations throughout the cross section, the flow is actually two or three-dimensional.

For spatial variation of flow rate along a sanitary sewer, the inflow is usually assumed to be concentrated at certain points. This assumption can also be used for points of outflow, but may not be needed since outflow in actuality tends to be concentrated.

In many instances it is sufficient to assume that the flow velocity is uniform across each section of flow. That is, the flow is assumed to be one-dimensional. While this assumption is adequate for most calculations, the variation of velocity with depth enters into such concerns as self-cleansing velocities in sewers.

Wastewater is essentially incompressible and, except for the possibility of the occurrence of water hammer in pressure conduits, incompressibility can be assumed.

Although the flow in sanitary sewers is essentially unsteady, significant changes in flow rate often require periods of hours. The dynamic effects of these relatively slow changes are negligible. (In open channels, the water surface slope for a given flow velocity may be the same as it would be for steady flow at that velocity.) Thus, the flow at any instant is treated as steady even though it varies significantly from hour to hour. This situation is often called quasi-steady flow.

In summary, most of the hydraulic calculations for sanitary sewers are based on the assumptions of one-dimensional, incompressible, steady flow with constant rate of flow between concentrated inflow or outflows. The calculations deal primarily with change in depth and velocity of flow along the sewer. Only occasionally do the calculations deal with two-dimensional or unsteady flow.

2. Examples of Flow Situations Encountered

While only a few basic types of flow are found in most sanitary sewers,

HYDRAULICS OF SEWERS

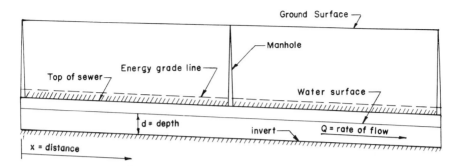

Depth d constant with distance x
Rate of flow Q constant with distance x

Fig. 5-1. Uniform conduit with steady uniform open-channel flow.

these occur in a variety of forms. It is worth a brief look at some specific examples of the forms in order to recognize situations where the basic types of flow occur.

The simplest type of flow is shown in the sanitary sewer profile of Fig. 5-1. In the reach shown, the conduit is of constant size and shape. The surface is open to the atmosphere, and the depth is constant with respect to the distance x along the sanitary sewer. The profile of the water surface is a straight line parallel to the invert. This is a case of open-channel flow referred to as steady uniform flow. Strictly speaking, the flow cannot be uniform unless the conduit is also uniform — that is, of constant size, shape and interior roughness. Nevertheless, the word "uniform" is usually applied to the flow. Even in a uniform conduit, this kind of flow does not exist unless the rate of flow Q is constant with respect to both distance x and time t.

Usually, the flow in a sanitary sewer cannot be uniform when there are inflows. Fig. 5-2 indicates the two basic types of inflow that can occur in a reach of a sanitary sewer. One type is a set of smaller inflows that occur at many points along the reach; in a smaller sanitary sewer these are from service connections, while for a larger sewer they may be from branches that are small relative to the size of the main sanitary sewer. The other type is the larger inflow that is concentrated at a small number of points along a reach; these are usually from branches that are large relative to the main sanitary sewer.

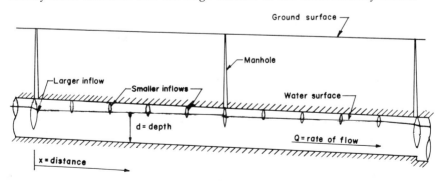

Depth d varies with distance x. Variation exaggerated for pictoral representation
Rate of flow Q varies with distance x.

Fig. 5-2. Typical water surface profile in sewer with lateral inflows.

When it is necessary to treat the flow as spatially varied, the smaller inflows can be approximated by an inflow that is distributed continuously along the reach. Distributed inflow is expressed in cubic meters per second per meter (cubic feet per second per foot) of distance along the sewer. The flow rate Q in the sewer is then a function of distance x as for steady flow.

If all of the inflows are constant with respect to time, the spatially varied flow is steady but not uniform. The depth or pressure upstream of an inflow must usually be greater than that downstream in order to cause an inflow to move in the direction of flow in the sanitary sewer and to accelerate to the velocity of that flow. Between inflows, however, the depth may increase in either the upstream or downstream direction depending upon the relationship between the flow rate and the slope of the conduit. The result is a complicated water surface, such as the one shown in Fig. 5-2.

In sanitary sewer design, the spatial variation of flow between the larger inflows is usually ignored. The total of the smaller inflows is added to the inflows at the end of the reach to form concentrated inflows. Between these inflows, the rate of flow Q is treated as constant with respect to x. Thus, the change in depth caused by the variation in Q is seldom taken into account.

While open-channel flow is most common in sanitary sewers, it is not universal. Fig. 5-3 shows an example of pressure-conduit flow. The water surface has risen up into the manholes, and the flow in the conduit is under pressure. Such flow occurs when the rate of flow exceeds the capacity of the sanitary sewer as an open-channel, when there is a serious obstruction in the sanitary sewer, or when a reach downstream has been forced to flow full for some reason. In some situations pressure flow may be temporary or intermittent. The capacity of sanitary sewers may be exceeded during peak rates of flow, but not during the lower rates. The discharge end of a sanitary sewer may be submerged only when the water surface in the receiving water or in a tank is at a high level. In other situations, the sanitary sewer may be designed for continuous pressure or surcharged flow.

Another common situation is a change in slope of the conduit. Part (a) of Fig. 5-4 shows an increase in slope with no change in conduit size. Since the downstream reach is steeper, the velocity V in that reach is greater. The change in the water surface at the change in slope depends on the amount of inflow at that location. If the inflow increases the flow rate to match the increased capacity of the downstream reach, the water-surface profile may simply follow the change in slope, as indicated by the broken line. On the

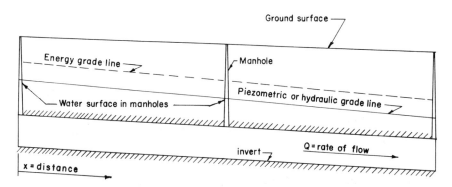

Fig. 5-3. Pressure conduit or pipe flow in sewer.

HYDRAULICS OF SEWERS 73

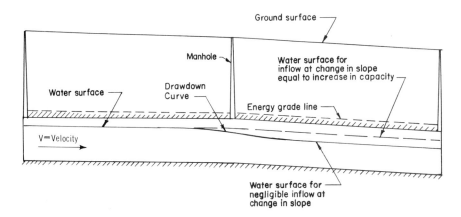

(a)

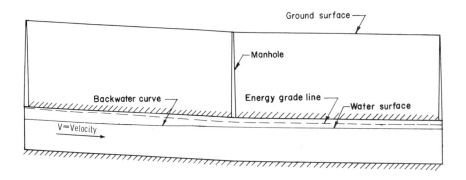

(b)

Fig. 5-4. Change in slope of conduit with no change in size: (a) increase in slope; (b) decrease in slope.

other hand, if the inflow is small or negligible, then the depth in the downstream reach must be less than that upstream. The depth in the upstream reach decreases toward the end of that reach, and the water-surface profile forms a drawdown curve that is convex upward. At the upstream end of the downstream reach, the water-surface profile may be either straight or concave upward.

Part (b) of Fig. 5-4 shows a change to a flatter slope with no change in size. The depth of flow in the downstream reach is greater than it is in the upstream reach, while the velocity is less. The greater depth backs up the flow in the upstream reach. Near the change in slope, the depth in the upstream reach increases with distance x, and the water-surface profile forms a backwater curve that is concave upward. A special case of this kind of change in slope is discussed below in connection with the hydraulic jump.

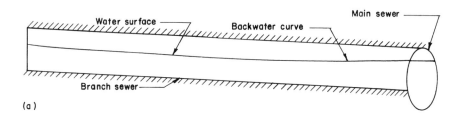

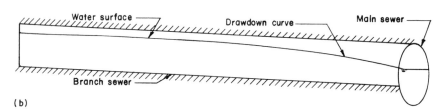

Fig. 5-5. Water surface profiles in branch sewer caused by flow in main sewer: (a) low flow in branch with peak flow in main; (b) peak flow in branch with low flow in main.

In sanitary sewers, drawdown and backwater curves are encountered in many other situations. Fig. 5-5 shows examples. Parts (a) and (b) show a profile of a branch joining a larger sewer. Due to the variation of flow with time, the flow depth in the branch may be low when the flow depth in the larger sanitary sewer is high, causing a backwater curve to be formed in the branch as a transition of flow surfaces. Conversely, a high flow in the branch with a low flow in the larger sanitary sewer produces the transitional drawdown curve in the branch. Similar situations can occur in a sewer that enters the forebay (wet well) of a sewage pumping station. A drawdown curve is also formed just upstream of a drop manhole or other transition structure.

Flow expansions occur whenever the size of the conduit is increased. Part (a) of Fig. 5-6 shows the sanitary sewer profile, the water-surface profile, and a horizontal section for an abrupt increase in sanitary sewer size. The broken lines below the water surface represent streamlines that indicate how the water is moving. In both the horizontal and vertical planes, the fluid cannot follow the conduit walls, and there is considerable eddy and turbulent motion before regular motion is established in the downstream conduit. The irregular motion results in a loss of flow energy at the expansion, which under certain circumstances can be significant.

Part (b) of Fig. 5-6 shows the same vertical and horizontal sections for a more gradual expansion at an increase in sanitary sewer size. The fluid can follow the conduit walls, as shown by the streamlines. As a result, the flow through the transition is smoother than it was in the abrupt expansion of Fig. 5-6 (a), and the loss in energy is much less.

Flow from a conduit into a tank is similar to flow expansion, but it is more extreme. An example is the flow into the forebay of a sewage lift or pumping station. Simplified plan views of this situation for abrupt and gradual connections between the conduit and the tank are shown in Fig. 5-7. In the abrupt connection shown in Part (a), the flow enters the tank as a jet whose energy must be dissipated in the tank. All of the kinetic flow energy that existed in the pipe is lost. In the gradual expansion in Part (b), due to the greater area of flow and consequent lowered velocities, the fluid slows down

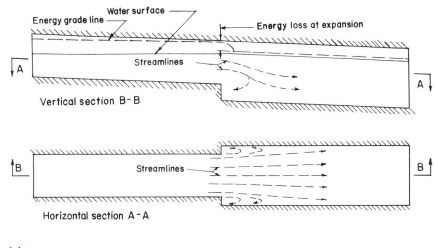

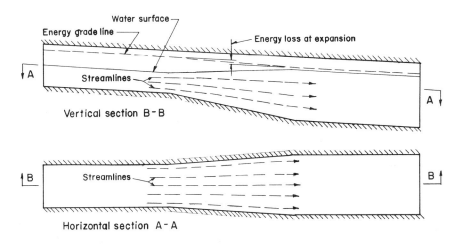

Fig. 5-6. Flow at increase in sewer size: (a) abrupt expansion; (b) gradual expansion.

before it enters the tank, and the amount of energy lost in dissipation is much less.

Flow through junctions occurs frequently in sanitary sewers. Fig. 5-8 shows the plan and profile of an example of such flow. The flow from the branch has components of velocity that are both perpendicular and parallel to the flow in the main conduit. Usually, the parallel component must be increased to match the velocity of flow in the main. For open-channel flow, this acceleration requires the water depth upstream of the junction to be greater than that downstream by an additional amount over and above the amount needed to overcome energy losses.

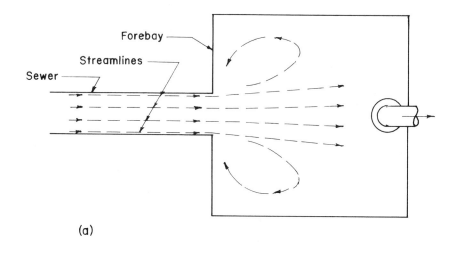

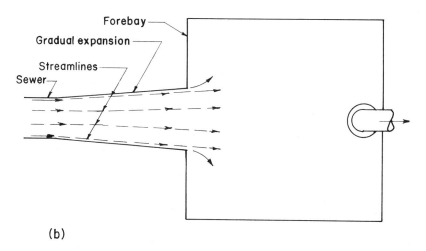

Fig. 5-7. Flow from sewer into tank or pump forebay: (a) abrupt outlet (plan view); (b) gradual expansion at outlet (plan view).

Another set of flow situations in sanitary sewers involves critical flow in open channels. Critical, subcritical and supercritical flow cannot be discussed adequately without some of the definitions and equations introduced below. However, the situations in which they are encountered can be described at this point.

Fig. 5-9 shows an example profile of a sanitary sewer at a change in slope from steep to mild. Upstream of the change, the steep slope produces a velocity that is greater than a certain critical value and a small depth of flow; the flow is called *supercritical*. For the same rate of flow, the mild downstream slope produces a velocity that is less than the critical value but with greater depth; the flow is called *subcritical*. Somewhere near the change in slope, the depth increases abruptly from the smaller depth to the greater depth to form a *hydraulic jump*. The jump itself takes place over a relatively short distance. It

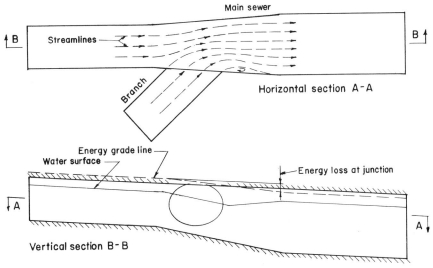

Fig. 5-8. Flow at junction.

has an irregular surface with a high degree of turbulent motion, mixing and energy dissipation. As shown further on in this chapter, the jump can occur on either the steep or the mild slope.

Careful consideration should be given to design situations where a hydraulic jump could form. The rapid decrease in flow velocity across the jump may encourage deposition of solids in the downstream conduit.

For many reasons the flow in sanitary sewers is basically unsteady. The inflows are seldom constant with respect to time. If the inflows are from service connections, they are intermittent and range in rate from zero to some maximum. Such inflows cause unsteady flow in the smaller sanitary sewers, which leads to unsteady flow in the whole sewer system. In addition, unsteady flow can result from the starting or stopping of pumps, or from the discharge from force mains. Since infiltration and inflow are seldom constant, they also contribute to unsteady flow. Sudden large releases to sanitary sewers from large industries is another cause of unsteady flow.

If any of these causes produces significant changes in flow rate within a period of seconds or minutes, then the dynamic effect of the change is important enough to change the slope of the water surface or to cause a drawdown when pumps are started. For open-channel flow, waves will be

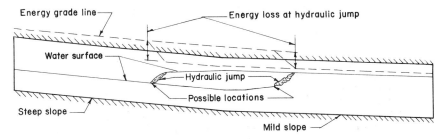

Fig. 5-9. Hydraulic jump at change in slope.

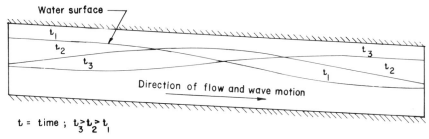

Fig. 5-10. Dynamic wave passing through reach of sewer.

formed on the water surface. These are called *dynamic waves* because the change in velocity is rapid enough to affect the slope of the water surface. They are also called *gravity waves*. Dynamic or gravity waves may move both upstream and downstream. Fig. 5-10 shows the water-surface profile at three consecutive instants of time as a dynamic wave passes through a reach of a sanitary sewer in the downstream direction.

As previously mentioned, significant changes in flow rate in sanitary sewers often require periods of hours rather than of minutes or seconds. The dynamic effects of such changes are negligible. In open-channel flow, the slope of the water surface for any flow situation is the same as it would be for steady flow in an otherwise identical situation. Such flow is quasi-steady. An important aspect of this is the change in channel storage that occurs when the water surface moves from one position to another. The most frequent cause of quasi-steady flow is the diurnal variation of the flow rate.

3. Equations of Motion

a. Analysis of One-Dimensional Steady Flow

The equations of motion for one-dimensional steady flow with constant flow rate along the conduit are simpler, special forms of the more general equations for one-dimensional, spatially varied unsteady flow. The general equations can be derived first and reduced to the special forms, but both the derivation and the general equations themselves tend to be complex. Over the years, the analysis of the simpler flow has evolved independently, and the simpler equations can be obtained and used directly without reference to the general equations. The simpler analysis is discussed and used in this chapter. However, the general equations must be used for more complex flows such as waves in open-channel flow (2,3,6,7,8), unsteady flow in pressure conduits (7,9), and water hammer (9,10). Recently, the calculations for many of these complex cases have been done on the digital computer (7).

b. Continuity Principle

The continuity principle is a statement of the conservation of mass, applied specifically to fluid flow. In the pressure conduit shown in Fig. 5-11, there is no inflow or outflow between Cross-Sections 1 and 2. There is no accumulation of fluid between the sections because the conduit is full and because water is treated as incompressible. Thus, inflow equals outflow. To express this equality let A represent the area of a cross-section, V the average flow velocity through the section, and Q the discharge (or volume rate of flow) through the section. These are related by:

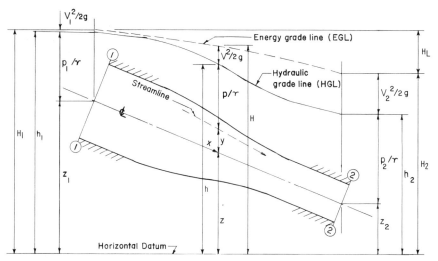

Fig. 5-11. Hydraulic and energy grade lines for pressure conduit flow.

$$Q = AV \tag{5.1}$$

Eq. 5.1 might be considered as the definition of the average velocity over the section or of the uniform velocity corresponding to the assumption of one-dimensional flow. If the subscripts 1 and 2 refer to Cross-Section 1 and 2, respectively, and if Eq. 5.1 is applied to both sections, the continuity principle can be expressed by:

$$A_1 V_1 = A_2 V_2 \tag{5.2}$$

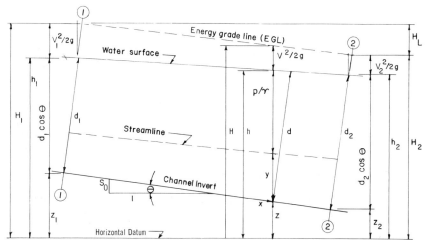

Fig. 5-12. Hydraulic and energy grade lines for open channel flow.

To apply the continuity principle to the open channel of Fig. 5-12, the area A must refer to the cross-section of the liquid flow, not the conduit. For steady flow the surface is stationary, and the volume below the surface is constant. Under these conditions, Eq. 5.2 applies to the open channel.

c. Energy Principle

At any point in a flowing fluid the total energy per unit mass is the sum of the potential energy, pressure energy, and kinetic energy. These three forms can be expressed in terms of the variables shown in Figs. 5-11 and 5-12. A typical flow streamline, which is a line that is everywhere parallel to the instantaneous direction of fluid motion, is shown for each conduit. The elevation of a point on the streamline above some horizontal datum is the sum of the elevation z of a point on the conduit plus the elevation y of the streamline above the point. This elevation is the potential energy per unit mass of fluid. The pressure energy per unit weight of fluid at the point on the streamline is equal to the pressure p at the point, divided by the fluid density γ, that is p/γ. Since this energy term has the units of length, it is often called the *pressure head*. The kinetic energy per unit weight is the *velocity head* $v^2/2g$, in which v is the local velocity along the streamline. The total energy, usually represented by H, can now be expressed as:

$$H = z + y + \frac{p}{\gamma} + \frac{v^2}{2g} \tag{5.3}$$

Since H is energy per unit weight, it has the unit of length, and it is often referred to as the *total head* or total *energy head*.

In one-dimensional flow, the local velocity v is equal to the average velocity V for all streamlines. Therefore, z and y can be referenced to any streamline. For the closed conduit, it is convenient to let z be the elevation of the centerline and let y be zero. Eq. 5.3 becomes:

$$H = z + \frac{p}{\gamma} + \frac{V^2}{2g} \tag{5.4}$$

For the open channel, it is convenient to let z be the invert elevation and let y be the depth. The term p/γ is then zero. Eq. 5.3 then applies to the open channel as well.

Two basic concepts or tools for the simpler equations and analysis are the hydraulic and energy grade lines. In Figs. 5-11 and 5-12 they are shown for pressure-conduit and open-channel flow, respectively. The *hydraulic grade line* (HGL) is a line showing the level to which water would rise in piezometers inserted into the flow at points along the streamline. The vertical distance between the HGL and the horizontal datum is the piezometric head, often called simply the head. The *energy grade line* (EGL) is a line showing the total energy of flow at points along the conduit in terms of equivalent elevation (or potential energy). The vertical difference between the two lines is equal to $V^2/2g$, the kinetic energy per unit weight of fluid. Energy grade lines are also indicated in some of the example flow situations in Figs. 5-1 through 5-10.

For the open channel shown in Fig. 5-12, the sum $y + p/\gamma$ of potential and pressure energy heads is not exactly equal to d (the depth) but to d (cos θ). When θ, the slope angle of the invert, is small enough for cos θ to be approximately one and the streamlines have negligible vertical acceleration and curvature, the sum of the two heads is equal to the depth. Eq. 5.3 can then be written as:

$$H = z + d + \frac{V^2}{2g} \tag{5.5}$$

In some important cases of open channel flow, the slopes of the invert are not important, and it is useful to use the invert of the conduit as the horizontal datum. The total energy, measured from this datum, is called the *specific energy* and is represented by H_0. The expression for specific energy is:

$$H_0 = d + \frac{V^2}{2g} \tag{5.6}$$

The difference in total energy (or total head) between cross sections in a conduit, is the *enegy loss* or *head loss* H_L. In Figs. 5-11 and 5-12, the loss between Sections 1 and 2 for the closed conduit is expressed as:

$$H_L = z_1 + \frac{p_1}{\gamma} + \frac{V_1^2}{2g} - z_2 - \frac{p_2}{\gamma} - \frac{V_2^2}{2g} \tag{5.7a}$$

The loss between Sections 1 and 2 for the open channel is expressed as:

$$H_L = z_1 + d_1 + \frac{V_1^2}{2g} - z_2 - d_2 - \frac{V_2^2}{2g} \tag{5.7b}$$

Eq. 5.7 is a form of the *energy equation*. A simlar equation can be written for specific energy if the slope of the invert between the two sections is zero. The energy equation is represented graphically by the energy grade line, and the energy loss can be represented graphically by a drop in the energy grade line, as shown in Figs. 5-11 and 5-12.

For some applications of the energy equation, the average velocity V is not an adequate approximation of the actual distribution of the local velocity v over the cross section. Specifically, there may be a significant error in the kinetic energy because the cube of the average velocity (V^3) is not equal to the average value of the cube of the local velocity (v^3). In such cases it is necessary to provide an energy coefficient α for the kinetic energy in Eqs. 5.4 through 5.7. Eq. 5.4 can then be written as:

$$H = z + \frac{p}{\gamma} + \frac{\alpha V^2}{2g} \tag{5.8}$$

while Eq. 5.5 becomes:

$$H = z + d + \frac{\alpha V_2}{2g} \qquad (5.9)$$

More information on the energy coefficient can be found in standard texts (2,5,6).

d. Momentum Principle

The relationship between force and acceleration contained in Newton's second law, can be expressed in a form suitable for one-dimensional flow in conduits. For the conduits in Figs. 5-11 and 5-12, the form is:

$$\sum F = \frac{\gamma}{g} Q (V_2 - V_1) \qquad (5.10)$$

in which $\sum F$ is the sum of all forces acting on the fluid contained between the two cross sections. These forces include the pressure forces pA at each section, the weight of the fluid, and the action of the conduit on the fluid necessary to constrain and guide the flow.

Eq. 5.10 is called the *momentum equation*. Since it is a form of the relationship between force and momentum, it is a vector equation, and the forces on the left side of the equation must be in the same direction as the velocities on the right. For one-dimensional flow, only forces and velocities in the direction of that flow are included in the equation. This concept is especially important for flow entering the conduit from the side and joining the flow in the conduit, as mentioned above in the discussions of lateral inflows and junctions.

For some applications of the momentum equation, the average velocity V is not an adequate approximation of the actual distribution of the local velocity v over the cross section. There may be a significant error in the flux of momentum through the cross section because the square of the average velocity (V^2) is not equal to the average of the square of the local velocity (v^2). In such cases it is necessary to provide a momentum correction factor β for the momentum in Eq. 5.10. The equation is then written as:

$$\sum F = \frac{\gamma}{g} Q [(\beta V)_2 - (\beta V)_1] \qquad (5.11)$$

4. Methods of Application and Calculation

a. Friction Losses

A large part of the hydraulic calculations for sanitary sewers is based on the continuity and energy equations. Essentially, the approach is to calculate the slope and drop of the energy grade line shown in Fig. 5-12.

In the space between the ends, inflows, controls or other changes in a conduit, the energy loss is the work done to overcome the force of frictional

HYDRAULICS OF SEWERS 83

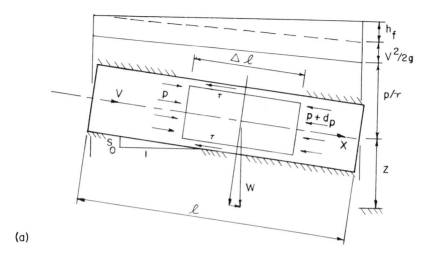

(a)

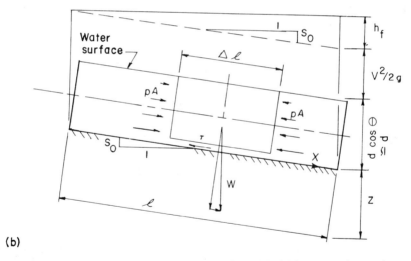

(b)

Fig. 5-13. Friction and other forces on fluid flowing in conduit: (a) forces on elemental length of fluid in pressure conduit flow; (b) forces on elemental length of fluid in open channel flow.

resistance between the water and the conduit. The expressions used to calculate the force and the energy loss are based on steady flow in pipes and steady uniform flow in open channels. The forces acting on elemental lengths of conduit for each of the two types of flow are shown in Fig. 5-13. These forces are the pressure p, the weight W, and the friction τ_0 between the fluid and the walls of the conduit. A balance of forces leads to equation 5.12 for pipe flow and Eq. 5.13 for open channels:

$$\tau_0 = \gamma R \frac{d}{dx}\left(\frac{p}{\gamma} + z\right) \qquad (5.12)$$

$$\tau_0 = RS_0 \qquad (5.13)$$

In these equations, R is the hydraulic radius. It is related to the area A and wetted perimeter P of the flow cross section by:

$$R = \frac{A}{P} \qquad (5.14)$$

For a circular closed conduit, R is equal to one-fourth of the diameter D.

It should be emphasized that Eqs. 5.12 and 5.13 involve no assumptions about friction factors or coefficients. They are merely an expression of a balance of forces. It should also be noted that Eq. 5.13 (for open channels) is based on steady uniform flow. In that case, both the water surface and the energy grade line are straight and parallel to the invert. Thus, the slope of the invert is used in the equation.

With the term τ_0, Eqs. 5.12 and 5.13 are not in a suitable form for calculating energy losses due to friction. Detailed analyses of two-dimensional flow has shown that the friction force is proportional to the square of the average velocity in both pipe flow and uniform open-channel flow. Thus, the friction term can be removed from the equations by substituting a term containing the square of the velocity and a suitable coefficient. The result is the well known Darcy-Weisbach equation for pipe flow:

$$\frac{h_f}{l} = \frac{fv^2}{2gD} \qquad (5.15)$$

and the well known Chezy equation for open channels:

$$V = C_f (RS_o)^{1/2} \qquad (5.16)$$

In Eq. 5.15 h_f is the loss of energy head in a uniform conduit of length l and diameter D, and f is the friction coefficient. In Eq. 5.16 C_f is the friction coefficient.

For calculation of energy losses due to friction, the Chezy equation must be rearranged in the form:

$$S_e = \frac{V^2}{C_f^2 R} \qquad (5.17)$$

The invert slope S_0 has been replaced by S_e, which is used to indicate the friction slope or slope of the energy grade line. The energy losses or head losses due to friction are then calculated by:

$$h_f = lS_e \tag{5.18}$$

in which h_f is the energy loss and l is the length of conduit for which the loss is calculated.

When open channel flow is steady and uniform, the slopes of the water surface and energy grade line are equal to the slope of the invert. For any discharge there is one depth at which such flow can satisfy Eqs. 5.16 or 5.17. The depth is often called the *normal depth* d_n, and the flow rate at that depth is often called *normal flow* Q_n.

Eqs. 5.15 and 5.17 are used to calculate the friction slope and friction losses for any one-dimensional flow, steady or unsteady, uniform or non-uniform. The assumption is that for any cross section of the conduit at any instant of time, the local energy losses due to friction for a given velocity and hydraulic radius are the same as would occur with steady pressure conduit flow or uniform steady open-channel flow having the same velocity and hydraulic radius.

The main problem in the use of these equations is the estimation of the coefficients f or C_f for a particular flow situation. The problem is addressed in Section D of this chapter.

b. Other Energy Losses

In addition to the effects of friction along the conduit, there is usually a local loss of energy associated with any sudden change such as an expansion or junction, with any flow control devices such as a weir or gate, or with flow between conduit and tanks. These losses occur over a relatively short distance along the conduit and are usually represented by a steep slope or sudden drop in the energy grade line. This kind of loss is associated with most of the flow situations represented in Figs 5-1 through 5-10, and the sudden change in the energy grade line is shown in some of these figures.

In terms of calculation, these losses are usually expressed either as loss or discharge equations, corresponding to the form of Eqs. 5.16 or 5.17, respectively. In the former the head loss ΔH_L can be expressed by:

$$\Delta H_L = K \frac{V^2}{2g} \tag{5.19}$$

in which K is a loss coefficient depending on the type of loss, and $V^2/2g$ is the velocity head. This form is normally used for losses due to flow in bends, expansions, contractions, from tanks into pipes, or out of pipes into tanks. The discharge equation is usually expressed as:

$$Q = C_q A \, (2g\Delta H)^{1/2} \tag{5.20}$$

in which C_q is a discharge coefficient, A is the area of the opening (such as the area under a sluice gate) and ΔH is the energy required to produce the discharge. This form is normally used for flow through nozzles, orifices, gates, and weirs. In order to express the loss, Eq. 5.20 must be rearranged in the form:

$$\Delta H_L = \frac{1}{2g}\left(\frac{Q}{C_q A}\right)^2 \tag{5.21}$$

The values of loss and discharge coefficients are found in many hydraulic texts and handbooks (2,3).

A third type of loss is usually determined from a combination of the energy and momentum equations rather than from a coefficient. This type of loss includes the hydraulic jump and the junction loss. Each of these is important enough to be discussed individually.

Historically, such losses have also been called *minor losses* because each one is usually small compared to the friction loss h_f along the conduit. With the flat slopes that many sanitary sewers have, that name may not be appropriate. For example, uniform flow at a slope of 0.001 requires a friction loss h_f of 1 m (3.28 ft) per 1,000 m (3,280 ft) of length. A velocity head of 0.1 m (0.33 ft) and a loss coefficient of 0.5 means a local loss of 0.05 m (0.16 ft). Only five such losses per 1,000 m would total 0.25 m (0.82 ft) or 25% of h_f. In sewer design, ignoring the local losses could lead to serious difficulties.

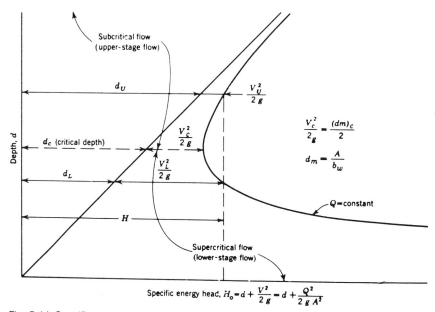

Fig. 5-14. Specific energy curve.

c. Specific Energy and Alternate Depths

For a given rate of flow and channel cross section, the specific energy H_o in Eq. 5.6 is a function of the depth. A plot of this function produces a specific energy curve like the one shown in Fig. 5-14. There is one depth at which H_o is a minimum. That is the *critical depth* d_c, and the corresponding velocity at that depth is the *critical velocity* V_c. Each larger value of H_o can occur at either of two *alternate depths*. The *upper depth* d_u, is greater than d_c, while the corresponding velocity V_u is less than V_c. This flow is *subcritical*. The lower depth d_L is less than d_c, with the corresponding velocity V_L greater than V_c. This flow is *supercritical*. The specific energy curve always has two asymptotes, as indicated in Fig. 5-14. One is the horizontal axis, $d = 0$, while the other is the 45° diagonal line, $H_o = d$.

An invert slope S_o on which the normal depth d_n equals the critical depth d_c is called a *critical slope*. A slope with d_n greater than d_c is a *mild slope*, while one with d_n less than d_c is a *steep slope*. Since d_n is determined by the hydraulic radius and resistance to flow, the hydraulic steepness of a channel slope is determined by more than S_o. A steep slope for a channel with a smooth lining could be a mild slope at the same flow for a channel with a rough lining. Even for a given channel, the slope may be mild for a low rate of flow and steep for a higher one.

By differentiation of H_o with respect to d, it can be shown (2) that the minimum value of H_o occurs when the velocity is related to the cross-sectional area A and width of water surface b_w by:

$$V_c = \left(\frac{gA}{b_w}\right)^{1/2} \tag{5.22a}$$

Since the ratio A/b_w is the hydraulic mean depth d_m, the critical velocity can be expressed as:

$$V_c = \sqrt{gd_m} \tag{5.22b}$$

The ratio of flow velocity to the critical velocity is the Froude number **F**. That is:

$$\mathbf{F} = \frac{V}{\sqrt{gd_m}} \tag{5.23}$$

The Froude number in Eq. 5.23 is the square root of the expression obtained from dimensional analysis. The latter is:

$$\mathbf{F} = \frac{V^2}{gd_m} \tag{5.24}$$

Eq. 5.23 is more convenient for analysis and is the expression used in this chapter.

In a rectangular channel, the relationship between minimum specific energy and critical depth is simple. The critical depth is equal to the mean depth and the velocity head is equal to $d_m/2$. Therefore:

$$(H_o)_{min} = \frac{3}{2} d_m \tag{5.25}$$

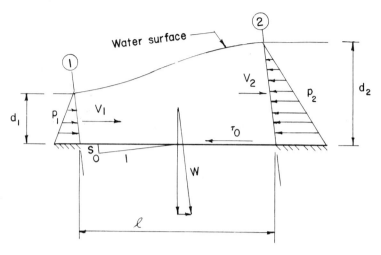

Fig. 5-15. Forces acting on fluid at hydraulic jump.

In open channels, many features of the flow and the water surface profile are determined by the Froude number and the value of the normal depth d_n relative to the critical depth d_c. These features are mentioned in Subsection **e** below.

d. Momentum Equation

The momentum equation is often used in situations where energy losses cannot be determined directly. For this purpose, Eqs. 5.10 or 5.11, the basic momentum equations, must usually be converted to forms more suitable for calculations. The expressions for momentum flux on the right-hand side are already suitable, but more explicit statements are required for the forces on the left-hand side.

Flow in a rectangular open channel leads to one of the simplest expressions for forces. Fig. 5-15 shows the forces acting on the fluid between two cross sections of an open channel. The pressure forces at Sections 1 and 2 are:

$$\left[\frac{\gamma \, (b_{w,1})(d_1)^2}{2} \right]$$

and:

$$\left[\frac{\gamma \, (b_{w,2})(d_2)^2}{2} \right]$$

respectively, in which b_w is the channel width. The component of weight along the channel is approximately:

$$\frac{\gamma l \, (A_1 + A_2) S_o}{2}$$

and the friction force is approximately:

$$\frac{\tau_o \, l \, (P_1 + P_2)}{2}$$

From Eqs. 5.13 and 5.17, the friction force can be expressed in terms of the friction slope as:

$$\frac{\gamma l \, (A_1 + A_2)(S_{e,1} + S_{e,2})}{4}$$

in which $S_{e,1}$, and $S_{e,2}$, are the value of friction slope at Sections 1 and 2, respectively. Substitution of these forces in Eq. 5.10 leads to:

$$\frac{1}{2}(b_{w,1}d_1^2 - b_{w,2}d_2^2) + \frac{l}{2}(A_1 + A_2)S_o - \frac{l}{4}(A_1 + A_2)(S_{e,1} + S_{e,2})$$
$$= \frac{1}{g}\left(\frac{Q_2^2}{A_2} - \frac{Q_1^2}{A_1}\right) \tag{5.26}$$

The momentum equation is often used in situations in which the length of open channel under consideration is short, or the invert and friction slopes are negligible compared to other terms in the equation. For such conditions in a rectangular conduit Eq. 5.26 can be reduced to the following form:

$$\frac{Q_1^2}{gA_1} + \frac{b_{w,1}d_1^2}{2} = \frac{Q_2^2}{gA_2} + \frac{b_{w,2}d_2^2}{2} \tag{5.27}$$

Eq. 5.27 suggests that a function can be defined by:

$$\mathbf{M} = \frac{Q^2}{gA} + \frac{b_w d^2}{2} \tag{5.28}$$

The function $\mathbf{M}$ has been called the *momentum function* (6) or *specific force* (2). Eq. 5.28 is equivalent to saying that the momentum function is constant between Sections 1 and 2.

With a slight modification, Eqs. 5.27 and 5.28 can be applied to non-rectangular conduits. The term $b_w d^2/2$ is replaced by the more general term $\overline{d}A$

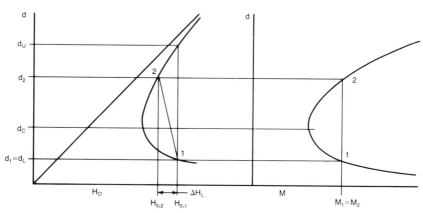

Fig. 5-16. Specific energy and momentum function at hydraulic jump.

in which $\bar{d}$ is the depth from the water surface to the centroid of area A. The momentum function becomes:

$$\mathbf{M} = \frac{Q^2}{2g} + \bar{d}A \qquad (5.29)$$

and the momentum equation becomes:

$$\frac{Q_1^2}{2g} + \bar{d}_1 A_1 = \frac{Q_2^2}{2g} + \bar{d}_2 A_2 \qquad (5.30)$$

It should be emphasized that this equation requires no assumption about the nature of the flow or the energy losses between the two sections.

For a given rate of flow the value of the momentum function varies with depth. In Fig. 5-16, for example, the momentum function is plotted in a form similar to the specific energy in Fig. 5-14. For comparison, the two are shown side by side. It can be shown (2) that the momentum function also has a minimum value at the critical depth d_c. For each value of **M** larger than the minimum, there are two *sequent* or *conjugate depths* d_1 and d_2 (11). The lesser depth d_1 is sometimes called the *initial depth*, while the greater depth d_2 is called the *sequent depth* (2).

The characteristic shape of the curve of the momentum function differs from that for specific energy. The momentum function has only one asymptote, the horizontal axis. The upper limb extends to the right without approaching the 45° diagonal.

The two curves shown in Fig. 5-16 help explain the height and energy loss for a hydraulic jump in a prismatic horizontal channel with no obstructions. In Fig. 5-15, the hydraulic jump can occur between Sections 1 and 2. Because of the complicated flow between the sections, the energy losses cannot be calculated directly. However, Eq. 5.30 indicates that the momentum function has the same value at both sections.

For the momentum function in Fig. 5-16, therefore, Points 1 and 2 are on the same vertical line. The initial depth d_1 upstream of the jump is the same as the lower depth d_L for specific energy. In the figure this equality is represented by the horizontal line between the two curves. Because of the shapes of the curves, however, the conjugate depth d_2 downstream of the jump is less than the alternate depth d_u. Thus, the value of $H_{o,2}$ of specific energy for Point 2 is less than the value $H_{o,1}$ for Point 1. The difference in depth $d_2 - d_1$ is the height of the jump, while the difference in specific energy $H_{o,1} - H_{o,2}$ is the energy loss H_L. The application of these concepts to flow in sanitary sewers is discussed below.

The hydraulic jump is only one example of the use of the momentum equations for calculations involving complicated flow patterns and energy losses. Another important example is flow at junctions. This situation usually requires a more complete form of the momentum equation as in Eq. 5.26. The simpler momentum function cannot be used because the flow rate changes between the two cross sections under consideration. The use of the momentum equation for flow at junctions was analyzed in detail by the Bureau of Engineering of the City of Los Angeles (12).

e. General Features of Water-Surface Profiles

In terms of water-surface profiles, open-channel flow can be categorized as gradually or rapidly varied. *Gradually varied* flow is based on the following assumptions (2):

(1) At a cross section of a conduit, the friction head loss is the same as it would be for uniform flow at that section with the same velocity and hydraulic radius.

(2) The slope of the conduit and vertical curvature are both small enough for the pressure in the flow to be hydrostatic.

(3) The channel is prismatic, i.e., of constant shape with constant alignment or small curvature.

(4) The momentum and energy correction factors are constant.

Rapidly varied flow is essentially all of the open-channel flow that is not gradually varied. Specifically, it has the following characteristics (2):

(1) The slope or curvature of the conduit is so pronounced that the pressure distribution in the flow is not hydrostatic.

(2) A rapid variation in flow occurs within a relatively short reach of conduit.

(3) Separation zones, eddies, and rollers may occur and complicate flow patterns or distort the velocity distribution.

(4) The energy and momentum correction factors vary greatly and may be difficult to determine.

For gradually varied flow, the water-surface profiles for a given flow rate are determined primarily by the relative magnitudes of the depth, the normal depth, and the critical depth. The last two are determined by the steepness of the slope, as already discussed. For mild slopes d_n is greater than d_c, while for steep slopes d_n is less than d_c. In addition there are critical slopes, adverse slopes and horizontal conduits. For each slope there are three basic kinds of profiles, depending on the ratio of the actual depth to d_n and d_c.

Fig. 5-17 shows the three basic types of profiles, M_1, M_2 and M_3, for mild slopes, and the three profiles S_1, S_2 and S_3, for steep slopes. Several of these have been mentioned above in the descriptions of situations encountered. A more complete listing of basic types with detailed discussion can be found in standard texts (2,3,6).

Rapidly varied flow is usually caused by an obstruction or change in the conduit. Often it is found at the end of a reach of gradually varied flow.

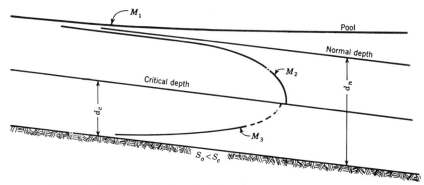

M_1 — Backwater from reservoir or from channel of milder slope $(d > d_n)$
M_2 — Drawdown, as from change of channel of mild slope to steep slope $(d_n > d > d_c)$
M_3 — Flow under gate on mild slope, or upstream profile before hydraulic jump on mild slope $(d < d_c)$

(a)

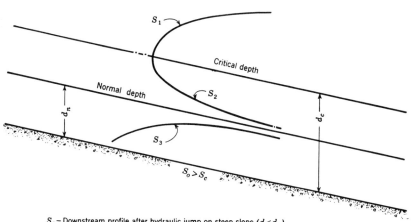

S_1 — Downstream profile after hydraulic jump on steep slope $(d < d_c)$
S_2 — Drawdown, as from mild to steep slope or steep slope to steeper slope $(d_c > d > d_n)$
S_3 — Flow under gate on steep slope, or change from steep slope to less steep slope $(d < d_n)$

(b)

Fig. 5-17. Open channel flow classifications: (a) mild slope $(d_n > d_c)$; (b) steep slope $(d_n < d_c)$.

Almost all flow over weirs, under gates, and through junctions or abrupt transitions is rapidly varied. The hydraulic jump is also rapidly varied flow. Compared to profiles for gradually varied flow, the water-surface profile for rapidly varied flow is often simply an abrupt change in depth or surface elevation. Usually the main task associated with it is to establish the location of the change and the water surface elevation on either side of it.

Fig. 5-18 shows an example of a combination of profiles for gradually and rapidly varied flow. The slope of the conduit changes from steep to mild, with supercritical flow in the upstream conduit and subcritical flow in the downstream conduit. To change from supercritical to subcritical, the flow must pass through a hydraulic jump with a consequent loss in head. For example,

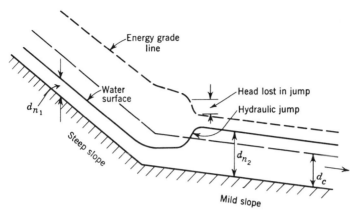

Fig. 5-18. Hydraulic jump profile.

in Fig. 5-18 the energy conditions are such that the jump occurs on the mild slope. If a greater downstream depth were necessary for the flow on the mild slope, then the jump would take place on the steep slope, as shown in Fig. 5-9.

The location and height of the jump are related to the profiles for gradually varied flow on either side of it. At some point downstream from the jump the depth and velocity of flow are determined. From that point a water-surface profile, either M_1 or M_2, is calculated and extended upstream toward the jump. The depth and velocity of flow must also be known at some point upstream on the steep slope. A water-surface profile, either S_2 or S_3, is calculated and extended from that point downstream toward the jump. Where these profiles overlap, the momentum function varies along each profile. The jump occurs where the momentum function has the same value on each profile. The height of the jump is the distance between the two profiles. The energy loss is determined by the method illustrated in Fig. 5-16. More detailed discussion of profile analyses and related calculations is found in standard references. (2,6).

D. FLOW RESISTANCE

1. Flow Friction Formulas

a. General

Eq. 5.16, the Chezy equation, was probably the first flow friction formula. It was developed by Antoine Chezy around 1768. The Chezy equation can be applied to both open-channel and pressure-conduit flow by using the appropriate expression for hydraulic radius and friction slope. The main requirement for using it is a suitable value or expression for the friction coefficient C_f. Numerous investigators have undertaken the task of determining how this coefficient varied under different conditions. An interesting summary of the various major investigations and formulas as well as an excellent bibliography are given by the American Society of Civil Engineers Task Force on Friction Factors in Open Channels (13). A critical analysis of open-channel resistance has been presented by Rouse (14). Present design practice tends to the use of the Manning formula for open-channel flow and the Manning, Hazen-Williams, or Darcy-Weisbach formulas for closed-conduit or pressure flow.

b. Kutter and Manning Formulas

Kutter's formula, published about 1869, received wide acceptance and use in estimating open-channel flows. The formula is rather unwieldly, but since it was accepted so widely, many tables and graphs for its solution were prepared. Through use of the formula, early designers became familiar with values of the Kutter roughness coefficient, n, applicable to sewers. Kutter's formula in metric units is:

$$V = \left[\frac{\frac{1}{n} + 23 + \frac{0.00155}{S_e}}{1 + \frac{n}{\sqrt{R}} \left(23 + \frac{0.00155}{S_e} \right)} \right] \sqrt{RS_e} \qquad (5.31a)$$

and in English units is:

$$V = \left[\frac{\frac{1.81}{n} + 41.67 + \frac{0.0028}{S_e}}{1 + \frac{n}{\sqrt{R}} \left(41.67 + \frac{0.0028}{S_e} \right)} \right] \sqrt{RS_e} \qquad (5.31b)$$

in which V is the mean velocity of flow, R is the hydraulic radius, S_e is the slope of energy grade line and n is the coefficient of roughness, and the bracketed portion of the equation is equal to C. Some authors have suggested that the term $0.00155/S_e$ ($0.0028/S_e$ in English units) be omitted since it originally was added to make the formula agree with data now known to be inaccurate (2).

The Manning formula has largely replaced the Kutter formula in engineering practice because of its greater simplicity and the fact that the n value is substantially equal to Kutter's n for types of pipe commonly used in sewer construction. The Manning equation is:

$$V = \frac{1}{n}(R^{2/3} S^{1/2}) \quad \text{(metric units)} \qquad (5.32a)$$

$$V = \frac{1.49}{n}(R^{2/3} S^{1/2}) \quad \text{(English units)} \qquad (5.32b)$$

in which the terms have been defined previously. The same value of n is used in both systems of units. Usual ranges of n for various conduit materials are listed in Table 5-1.

Table 5-1 Values of Manning Coefficient for Various Materials

Conduit Material (1)	Manning n (2)[a]
Closed conduits	
Asbestos-cement pipe	0.011-0.015
Brick	0.013-0.017
Cast iron pipe	
Cement-lined & seal coated	0.011-0.015
Concrete (monolithic)	
Smooth forms	0.012-0.014
Rough forms	0.015-0.017
Concrete pipe	0.011-0.015
Corrugated-metal pipe	
(½-in. × 2⅔-in. corrugations)	
Plain	0.022-0.026
Paved invert	0.018-0.022
Spun asphalt lined	0.011-0.015
Plastic pipe (smooth)	0.011-0.015
Vitrified clay	
Pipes	0.011-0.015
Liner plates	0.013-0.017
Open channels	
Lined channels	
a. Asphalt	0.013-0.017
b. Brick	0.012-0.018
c. Concrete	0.011-0.020
d. Rubble or riprap	0.020-0.035
e. Vegetal	0.030-0.40[b]
Excavated or dredged	
Earth, straight and uniform	0.020-0.030
Earth, winding, fairly uniform	0.025-0.040
Rock	0.030-0.045
Unmaintained	0.050-0.14
Natural channels (minor streams, top width at flood stage < 100 ft)	
Fairly regular section	0.03-0.07
Irregular section with pools	0.04-0.10

[a]Dimensional units contained in numerical term in formula.
[b]See References 2, 5, 16. (Varies with depth and velocity.)
Note: 1 in. = 2.54 cm; 1 ft = 0.305 m

    The Kutter and Manning equations are used for pipes and conduits of all shapes flowing either full or partly full. The graphs that are available for their solution have been compiled for full-conduit flow only, and the so-called hydraulic-elements graphs described below (or tabular values thereof) are used to determine the flow characteristics at other than full flow. Fig. 5-19 is a nomograph for the solution of the Manning equation for circular pipes flowing full, in metric or English units. This chart, as most others for the same

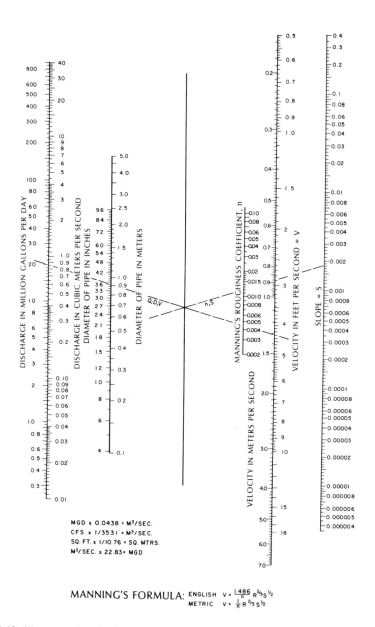

Fig. 5-19. Alignment chart for flow in pipes, Manning formula.

purpose, also provides a solution of the equation $Q_f = A_f V_f$ for full pipe. An example of the use of the chart is illustrated by the broken lines.

The chart can be used for other shapes of closed conduits and open channels if the discharge scale is ignored and the diameter scale is taken to represent values equal to four times the hydraulic radius of the actual cross section. If, for example, Q is given with depth to be determined, then the choice of S and n will indicate a point on the transfer line. By pivoting from this point, sets of values of $4R$ and V may be obtained. By successive trials, a compatible set of values for the given Q and the cross section may be found which furnishes the correct depth of flow, d.

For frequent use, graphs of slope vs. discharge, plotted on log-log scales with lines of constant conduit size and velocity, are probably more convenient than the nomographs of Fig. 5-19. Individual graphs are usually prepared for each n value. Such graphs may be found in many texts and in other source material.

Many engineers prefer to use $Q/S^{1/2}$ values (2,3) for closed-conduit flow; these values are constant for any given conduit size and n value. For open-channel flow, tabular values of $Qn/(d^{8/3}S^{1/2})$ are often used for design to determine depth of flow or channel size (3,17). Special slide rules are available for the solution of the Manning equation.

Alignment charts, graphs and tables for the Manning equation are beginning to be replaced by small digital computers and programmable calculators. With these devices, Eq. 5.32 can be solved directly for V, R or S if the values of the remaining two are known (18).

c. Hazen-Williams Formula

The Hazen-Williams formula is:

$$V = 0.85CR^{0.63}S^{0.54} \text{ (metric units)} \qquad (5.33a)$$

$$V = 1.32CR^{0.63}S^{0.54} \text{ (English units)} \qquad (5.33b)$$

in which the nomenclature is basically the same as that used in Eq. 5.31 and in which C is a coefficient. The formula is widely used for pressure-conduit or pipe flow, although it is equally applicable to open-channel conditions. Published values for C have come largely from pipe-flow experiments, and the values have been found to depend primarily on the roughness of the conduit.

While a wide variety of values that have been published in table form are available (19), the values typically used are between 100 and 140. The following values of C are suggested by Brater and King (3) for pipes carrying water. The C value for new, smooth pipes generally is taken to be between 130 and 140. For old, tuberculated cast-iron pipes, the C value could be 60 to 80, and for small pipes that have been badly tuberculated, the value could be as low as 40. Low C values resulting from tuberculation of cast iron pipe may be avoided by the use of cement-lined pipe. To estimate fraction losses for future conditions, lower values of C are used to allow for reductions in carrying capacity resulting from films and slimes on the pipe interior. For smooth concrete and cement-lined pipes a C value of 100 to 120 is commonly used for future conditions. Fig. 5-20 is an alignment chart in metric and English units for the solution of the Hazen-Williams formula for pipes flowing full. As mentioned previously for the Manning formula, graphs for slope vs. discharge plotted on log-log scales or special slide rules are probably more convenient for frequent use. Values of $Q/S^{0.54}$ can be computed and used for closed-conduit flow. However, all of these graphical aids are beginning to be replaced by small digital computer and programmable calculators. Eq. 5.33 can be solved for any one of the variables if values for the other two are known.

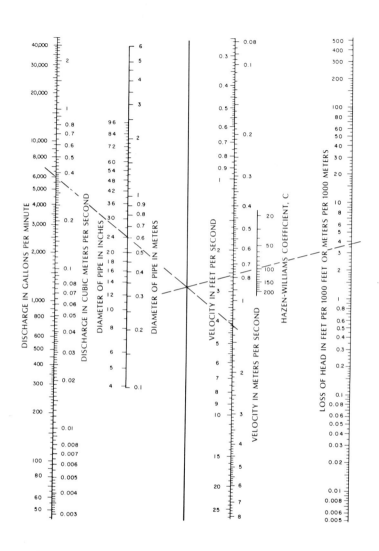

HAZEN-WILLIAMS FORMULA: ENGLISH V in fps $= 1.318\ C\ R^{0.63} S^{0.54}$
METRIC V in mps $= 0.8493\ C\ R_m^{0.63} S_m^{0.54}$

Fig. 5-20. Alignment chart for flow in pipes, Hazen-Williams formula.

d. Darcy-Weisbach Equations

Of all of the formulas currently in use, the Darcy-Weisbach equation, Eq. 5.15, is often called the most scientific because it maintains the basic form of the Chezy equation, Eq. 5.16. The coefficient f is dimensionless and it is based on the most scientific data. However, those data seldom correspond to the flow situation in sanitary sewers.

The coefficient f has been found to be primarily a function of the

Reynolds number and the relative roughness of the conduit. The Reynolds number, which is dimensionless, is defined normally as:

$$\mathbf{R} = \frac{4RV}{\nu} \qquad (5.34a)$$

in which ν is the kinematic viscosity. For a circular closed conduit flowing full, this reduces to:

$$\mathbf{R} = \frac{DV}{\nu} \qquad (5.34b)$$

The relative roughness is defined as the ratio of the height, k, of the roughness elements on the conduit walls to the size of the conduit. For circular pressure conduits, the relative roughness is k/D, and for open channels it is $k/4R$. The value of k is sometimes called the effective absolute roughness.

The magnitude of the Reynolds number determines whether the coefficient f depends more on the roughness or on the effects of fluid viscosity. At low Reynolds numbers there is a thick layer of a laminar viscous flow at the conduit wall. This layer covers more of the roughness, with less of the roughness elements protruding through it. Therefore, the wall seems smooth to the main flow beyond the laminar layer. At high Reynolds numbers the layer of laminar viscous flow is thinner and covers less of the roughness. Since more of the roughness elements protrude through the layer, the wall seems rougher to the main flow beyond the layer.

At some large Reynolds number, the laminar layer reaches a minimum thickness and does not change with additional increases in Reynolds number. The turbulence in the flow is said to be complete or fully developed. Because the two effects are concurrent, the conduit is sometimes said to be *rough* in the range of large Reynolds number with *complete* or *fully developed turbulence*. Of course, the conduit has not changed physically from smooth to rough.

A considerable amount of theory and experiment has been devoted to establishing the complex relationship between f, R, and k/D. Measurements have been made with artifically roughened pipe and commercial pipe, both of metal and concrete. The roughness of the conduit has been found to depend not only on the size of the pipe, but on the shape and spacing of the surface irregularities. For most commercially available pipe and job-formed conduits of materials similar to those used in commercial pipe, k is found to be relatively constant for a given material, and the relative roughness is adequate to define the conduit roughness.

For design purposes, the available information has been compiled into graphical forms for both pipes and open channels (2,5,20,21). The available information has also been presented in the form of equations expressing f as a function of **R** and k/D (22). In the past, such equations were not used extensively because of their complicated form. More recently, the equations have been programmed for small digital computers and programmable calculators (18). The availability of these devices to the individual engineer may lead to more use of the equation.

Several items work against the use of the Darcy-Weisbach equation for general design of sewers and open channels. From the significance of relative roughness and Reynolds number, it is reasonable to expect the friction coefficient to change with conduit size if the velocity of flow and size of roughness are held constant. Thus, the coefficient will vary with conduit size for a given conduit material and velocity. For fully developed turbulent flow in pipes flowing full (so-called fully rough pipes), it has been shown that the coefficient f is proportional to $(k/D)^m$, in which m is slightly less than 1/3 (6).

When applied to pipes flowing full, the Manning coefficient varies with conduit size. Comparison of the Manning and Darcy-Weisbach equations shows that n is proportional to $D^{1/6}f^{1/2}$. Since the Manning equation applies only to fully developed turbulence, the expression $(k/D)^m$ can be substituted for f. It follows then that n varies according to:

$$n \propto k^{m/2} D^{(1-3m)/6} \qquad (5.35)$$

Since m is slightly less than 1/3, n varies slightly with diameter D. Some studies have shown the percent variation in n to be less than 1/5 of that in f (3).

For open channels, the values of f are further complicated by the shape of the channel and the Froude number (1,3). Also to be considered is the state of the inside surface of the sewer after it has been used for some time. Deposits, surface erosion, and accumulated slime layers make the absolute and relative roughnesses difficult to define. These are all discussed below as factors affecting friction coefficients.

Considering all of these items, the Darcy-Wesibach equation, with its cumbersome friction coefficient, does not yet offer any better precision than the simpler Manning equation. The extra time that would be required to use the former would be better spent examining the flow situations and water-surface profiles that are anticipated.

2. Factors Affecting Friction Coefficients

a. General

There have been many experiments in the laboratory and the field to determine friction coefficients for various materials and conditions. Accurate measurements can be obtained in the laboratory, but it is difficult to duplicate conditions of flow that are equivalent to those in a sewer. On the other hand, field measurements in existing sewers may reflect unknown variables peculiar to the particular sewer being investigated, as well as errors in measurement and an inability to control identifiable variables.

Factors that affect the choice of a coefficient are the conduit material, size, shape, and depth of flow. In addition to these interrelated factors, the following should be considered:

- Rough, opened, or offset joints, and joint compounds;
- deposits in sewers, particularly grit accumulation in the invert;
- coatings of grease or other matter on interior of sewer;
- tree roots and other protrusions;
- flow from laterals disrupting flow in the sewer;
- typical slime layer growth on wetted surfaces that occurs in nearly all sewers.

b. Conduit Material

The material of which a conduit is constructed and the type of joint determine the interior surface of the conduit. Thus, they are the basic factors contributing to the roughness and resistance to flow. For the Manning and Hazen-Williams equations, the friction coefficient is selected directly on the basis of the material and joints. Table 5-1 is a summary of commonly used values of n compiled from the various sources. The values are for sewer design and hence are higher than the values obtained in laboratory tests with clear water and clean conduits.

The range in coefficients for a given pipe material is explained partially by the effect of disturbing influences that were mentioned in the general discussion of coefficients. Generally, Manning's n for a given sewer, after some time in service, will approach a constant which is not a function of the pipe material but represents the grit accumulation and slime build-up on the pipe walls. This n will be on the order of 0.013. A coefficient which will yield higher friction losses should be selected for sewers where disturbing influences are known or anticipated. Because of the empirical nature of each formula, conservative design is prudent.

c. Size of Conduit

The significance of relative roughness, Reynolds number, and conduit size were discussed in connection with the Darcy-Weisbach equations. Both the Manning and Darcy-Weisbach friction coefficient are found to vary with conduit size for a given conduit material and flow velocity. The variation of f often has been several times the variation in Manning n in the same situation.

d. Conduit Shape

The shape of the conduit has an effect on the coefficient in open channel flow. For a given degree of roughness, the coefficient f decreases approximately in the order of rectangular, triangular, trapezoidal, and circular conduits (2). The coefficient n varies in the same manner, but proportional to $f^{1/2}$. The effect of channel shape may be due to the existence of a component of flow velocity perpendicular to the longitudinal direction of the conduit. This perpendicular motion is called *secondary flow,* and it increases the energy loss during flow.

e. Depth of Flow

The effects of flow depth on f and n have not been completely defined. For closed conduits flowing part full, f and n have been observed to vary with depth of flow with no apparent change in boundary roughness. This variation may be a combination of the effects of size and shape already discussed.

Experiments (23,24) show that the value of n for a pipe flowing partly full is greater than for the same pipe flowing full. The average results are shown in Fig. 5-21, which is the hydraulic element graph prepared by Camp (25) for a circular conduit with uniform roughness over the entire inside surface. In the ratio n/n_f, n is the value for the partly full pipe, while n_f is the value for the full pipe. This ratio is shown as a function of the relative depth d/D. Therefore, reliability of the average is subject to question.

The ratio f/f_f is also shown in Fig. 5-21 by a curve and small circles. The value f is for the partly full pipe, while f_f is the value for the full pipe. From the discussion above of the Darcy-Weisbach equations, it can be seen that the ratios n/n_f and f/f_f are related by:

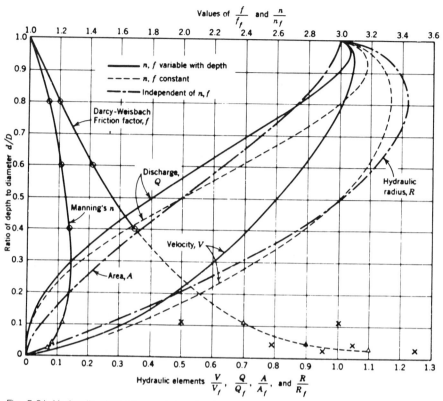

Fig. 5-21. Hydraulic elements graph for circular sewers.

$$\frac{n}{n_f} = \left(\frac{R}{R_f}\right)^{1/6} \left(\frac{f}{f_f}\right)^{1/2} \tag{5.36}$$

in which R is the value of the hydraulic radius for the partly full pipe and R_f is the value for the full pipe.

The points in Fig. 5-21 marked by triangles and x's were estimated from measurements made in Louisville, Ky. (26) in large sewers flowing part full. Tests on a large concrete arch-section sewer in Kansas City, Mo. (27) indicated a variation in n with depth, with n increasing with depth up to about one-half of the section height. Limited tests on asbestos-cement, cast iron, and vitrified clay pipes (28) apparently indicate agreement with the variation in n shown in Fig. 5-21. Recent tests of existing sewers (29) also indicate agreement with the variation in n for the larger depths of flow. At very low depths of flow the results of this experimentation indicate quite low n values, provided the sewers were clean and well constructed. It is interesting to note, however, that tests on hydraulically smooth, polyvinyl chloride pipe (30) flowing at relative depths of 0.2 to full do not indicate a significant variation in n with depth, and the data for discharge and velocity approximate the curves which are obtained for the assumption of constant n.

E. DESIGN COMPUTATION

1. Application of Hydraulic Computations

For sanitary sewer systems, hydraulic computations are used in planning, design, and interpretation of flow measurements. In planning or preliminary design, the computations are usually concerned with conduit capacity and maximum or minimum slopes. Typical design computations involve capacity, slope, and significant water-surface elevations at important locations in the conduit, such as upstream and downstream of junctions or changes in conduit size. It is often possible to assume that the water-surface profile between these locations is a straight line connecting the significant elevations. Occasionally, design requires more detailed computation of actual water-surface profiles, such as drawdown and backwater curves.

Only rarely does design involve such elaborate computations as routing unsteady flow and analysis of wave motion. However, these computations are more commonly becoming necessary for adequate interpretation of data on wastewater flows. It is often necessary to calculate the effects of backwater and drawdown as well as unsteady flow, in order to adjust the measurement data and obtain satisfactory results.

This section relates the situations, concepts, and principles discussed in previous sections to some of the hydraulic computations required in planning, design and flow measurements. Most of the computations and applications are only discussed or described, with no intent to provide sufficient detail for carrying them out. Such details are available from a variety of hydraulics texts, published articles, agency manuals, and other sources.

2. Capacity and Flow Estimates

a. Full Pipes

Many design computations begin with an estimate of the capacity of a pipe of a given size and roughness flowing full at a given slope, or with the selection of the size and slope of a pipe to carry a given discharge. Either Eqs. 5.32 or 5.33 or alignment charts (Figs. 5-18 and 5-20) are usually employed, and the solution for a full pipe is obtained. An example of the use of the alignment chart is shown in Fig. 5-19.

b. Partly Full Pipes

In order to consider the range of discharge that will occur in a partly full pipe it is often necessary to determine the depth and velocity of flow at discharges less than capacity. The partly full flow is assumed to be steady and uniform, and the depth and velocity are determined from a hydraulic element chart. For convenience, the velocity or flow rate in the partly full pipe can be expressed as a fraction of the corresponding value in the full pipe. Let a symbol without a subscript represent the value of a variable when the conduit is partly full and let the subscript "f" indicate values for the full conduit.

The flow ratio can be expressed as:

$$\frac{Q}{Q_f} = \frac{AV}{A_f V_f} = \left(\frac{n_f}{n}\right)\left(\frac{A}{A_f}\right)\left(\frac{R}{R_f}\right)^{2/3} \qquad (5.37)$$

The ratios Q/Q_f, V/V_f, A/A_f, R/R_f and n/n_f are called the hydraulic elements of the conduit. They are usually determined as functions of the relative depth

d/d_f and are often presented graphically as a hydraulic element chart or graph. The calculation of A/A_f and R/R_f from the geometry of the circle is described by Fair et al (31). These ratios are used in Eq. 5.37 with the ratio n/n_f (already discussed) to obtain the ratios V/V_f and Q/Q_f. Camp's hydraulic element chart for circular pipe (25) is shown in Fig. 5-21. Similar charts have also been developed for other conduit shapes (32,33). In Fig. 5-21, two sets of curves are shown for the velocity and discharge ratios. The broken lines show the values based on constant n values. For all d/D values the value of n is equal to n_f. The solid lines show the values that result if n varies as indicated by the curve for n/n_f.

Until more information and better analyses are available, the decision to use a constant or variable n must be left to the engineer. However, some aspects of the method of sanitary sewer design have a bearing on the decision. If the sanitary sewer is designed to carry the design flow when flowing full, the difference is less important. On the other hand, if the design flow is to be carried at a relative depth between 0.8 and 1.0, then the use of constant n might lead to a design flow rate that is greater than Q_f. If a flow situation caused the pipe to fill, then the pipe capacity would be less than the design flow. For either constant or variable n, it is unwise to have a design flow rate greater than Q_f.

The method of using the hydraulic-elements graph is illustrated below:

Example 5-1. A 300-mm (12-in.) sanitary sewer is laid on a slope of 3.0 units per 1,000 units. Find the velocity and discharge if the sewer is flowing 0.4 full (120-mm — 4.8-in. — depth). Assume Manning's $n_f = 0.013$.
   (a) From Fig. 5-19, $Q_f = 0.0566$ m³/sec (2.0 cfs), $V_f = 0.76$ m/sec (2.5 fps)
   (b) From Fig. 5-21, for $d/D = 0.4$,
     (1) n constant:
       $Q/Q_f = 0.33$; $V/V_f = 0.90$
       $Q = 0.33 \times 0.0566 = 0.0187$ m³/sec or 18.7 L/sec (0.66 cfs);
       $V = 0.90 \times 0.76 = 0.68$ m/sec (2.25 fps)
     (2) n variable:
       $Q/Q_f = 0.27$; $V/V_f = 0.71$
       $Q = 0.27 \times 0.0566 = 0.0153$ m³/sec or 15.3 L/sec (0.54 cfs);
       $V = 0.71 \times 0.76 = 0.54$ m/sec (1.78 fps)

Many engineers prefer a general tabulated version of the curve for Q/Q_f. With the Manning equation the ratio for circular conduits can be written as:

$$\frac{Q}{Q_f} = \frac{Q n_f (4)^{5/3}}{\pi D^{8/3} S^{1/2}} \quad \text{(metric units)} \tag{5.38a}$$

$$\frac{Q}{Q_f} = \frac{Q n_f (4)^{5/3}}{1.486 \pi D^{8/3} S^{1/2}} \quad \text{(English units)} \tag{5.38b}$$

It follows that:

$$\frac{Q n_f}{D^{8/3} S^{1/2}} = 0.3117 \frac{Q}{Q_f} \tag{5.39a}$$

and:

$$\frac{Qn_f}{D^{8/3} S^{1/2}} = 0.4632 \frac{Q}{Q_f} \tag{5.39b}$$

in metric and English units, respectively. The numerical value of the right-hand side of Eq. 5.39 can be tabulated as a function d/D as has been done for English units in some standard works (3,17). With such tables Q can be found for any value of d/D if D, S and n_f are known.

c. Self-Cleansing Velocities

Of particular concern in sanitary sewers is the deposition of suspended material. Particles that reach the bottom of the conduit do not remain as a deposit if the velocity and turbulent motion are sufficient to resuspend them or move them along the bottom. A velocity sufficient to prevent deposits is called *self-cleansing velocity*.

The velocity required to transport sediment in a pipe flowing full is (34,35):

$$V = \frac{R^{1/6}}{n} \left[B(s - 1)D_g \right]^{1/2}$$

$$= \left[\frac{8B}{f} g(s - 1)D_g \right]^{1/2} \quad \text{(metric units)} \tag{5.40a}$$

$$V = \frac{1.486}{n} R^{1/6} \left[B(s - 1)D_g \right]^{1/2}$$

$$= \left[\frac{8B}{f} g(s - 1)D_g \right]^{1/2} \quad \text{(English units)} \tag{5.40b}$$

where s is the specific gravity of the particle; D_g is the diameter of the particle; B is a dimensionless constant with a value of about 0.04 to start motion of clean granular particles and of about 0.8 for adequate self-cleansing of cohesive material; and the other terms are as previously defined.

Self-cleansing velocities at low flows when the conduit is only partly full should be evaluated. Fair et al (31) extended Camp's work by developing an equation for the slope of a partly full pipe that produces the same cleansing effect as when that pipe is flowing full. Fig. 5-22 gives the hydraulic elements of circular sewers that possess equal self-cleansing effect.

Fig. 5-22 can also be used for selecting pipe slopes for a given degree of self-cleaning when part full. Use of Eq. 5.40, with Eq. 5.32 or Fig. 5-19, enables selection of a pipe to carry a design flow Q_f at a velocity V_f that will move a grain of a given size D_g. This same particle is to be moved by lesser flow rates between Q_f and some lower value Q_s. For the ratio Q_s/Q_f, Fig. 5-22 shows the ratio S_s/S_f to move the same particles as well as the velocity ratio V_s/V_f and the relative depth d/D.

The figure shows that any flow ratio Q/Q_f that causes the relative depth

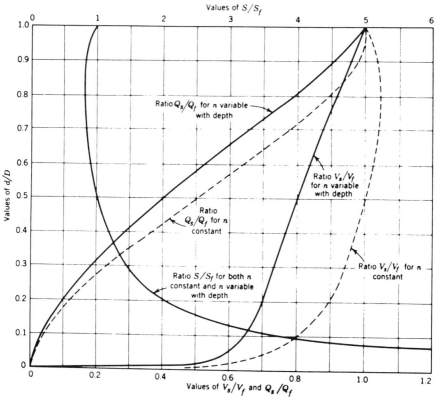

Fig. 5-22. Hydraulic elements of circular sewers that possess equal self-cleansing properties at all depths.

d/D to be greater than 0.5 requires no increase in slope because S_s is less than S_f. For smaller flows, the invert slope S_o must be increased to S_s to avoid a decrease in self-cleaning.

Example 5-2. A 25.4-cm (10-in.) sewer is to discharge 0.0113 m³/sec (0.4 cfs) at a velocity equivalent in self-cleansing action to that of the same size sewer flowing full at 0.61 m/sec (2.0 fps). Find the velocity of flow and the required slope for $n_f = 0.013$.

(a) From Eq. 5.40, with B = 0.8, grit of up to 0.12 mm (0.005 in.) will be transported effectively in a 25.4-cm (10-in.) sewer flowing full at 0.61 m/sec (2 fps). From Fig. 5-19, $Q_f = 0.031$ m³/sec (1.1 cfs) and $S_f = 2.5$ units per 1,000 units; hence, $Q_s/Q_f = 0.0113/0.031 = 0.364$.

(b) From Fig. 5-22 for constant n and $Q_s/Q_f = 0.364$, $V_s/V_f = 0.97$ and $S/S_f = 1.20$; hence $V_s = 0.97 \times 0.61 = 0.59$ m/sec (1.94 fps) and $S = 1.20 \times 2.5 = 3.0$ per 1,000. Thus, to effectively transport grit of up to 0.12 mm (0.005 in.), the sewer must be sloped at 0.003 if n is assumed constant.

(c) From Fig. 5-22 for variable n and $Q_s/Q_f = 0.364$, $V_s/V_f = 0.79$ and $S/S_f = 1.05$; hence, $V_s = 0.79 \times 0.61 = 0.48$ m/sec (1.58 fps) and $S = 1.05 \times 2.5 = 2.63$ per 1,000.

Where it is not feasible to conduct an analysis as in Example 5-2, a minimum velocity is often accepted as the design criterion. One minimum

Table 5-2 Slopes for Half-Full and Full Sewer Pipe Velocities of 0.6 m/sec (2.0 fps); N = 0.013

Pipe Diameters		Slope
in millimeters (1)	in inches (2)	(3) [meters per 100 m (feet per 100 ft)]
150	6	0.50
200	8	0.40
250	10	0.28
300	12	0.22
375	15	0.15

Note: For a given flow and slope, velocity is influenced very little, if at all, by pipe diameter. Calculations from Camp's curve of the hydraulic elements of circular section pipes (Fig. 5-24) and others show less variation than calculations based on classical equations.

often recommended is 0.61 m/sec (2.0 fps) (35,32,33). For circular conduits designed on this basis, a velocity of 0.3 m/sec (1.0 fps) occurs when the relative depth is about 0.22 and the flow rate is about 9% of the sewer capacity. The velocity of 0.3 m/sec (1.0 fps) has been found sufficient to prevent serious deposition of the lighter sewage solids, but serious deposition of sand and gravel is possible.

It is desirable to have a velocity of 0.91 m/sec (3.0 fps) or more whenever possible. Slopes corresponding to full-pipe velocities as low as 0.46 m/sec (1.5 fps) have been used successfully, but sewers at such slopes must be designed and constructed with more consideration.

Once a minimum velocity for full conduits has been selected, it is possible to determine the corresponding minimum slope for each size and roughness (35). As an example, Table 5-2 provides sets of slopes corresponding to a minimum velocity of 0.61 m/sec (2.0 fps) and Manning n of 0.013.

3. Water-Surface Elevations at Important Locations

a. Critical Depth

Critical depth frequently occurs in design computations. Normal depth is compared with critical depth to determine whether flow is supercritical or subcritical. At a drop in the invert or at a significant increase in conduit size, critical depth may be assumed and used as a starting depth for calculation of a drawdown curve.

In a rectangular conduit, critical depth can be determined easily by Eq. 5.25, but in conduits of other shapes, the calculation is more difficult. With the increased availability of small computers and programmable calculators, trial-and-error solutions for a set of d, A, and R values that satisfy Eqs. 5.32 and 5.25 (18) may be used.

In order to avoid trial-and-error computations, various tables, graphs, and alignment charts have been prepared for many conduit shapes. Fig. 5-23 shows an example alignment chart for circular pipes. An example calculation to demonstrate use of the chart is indicated by the broken lines.

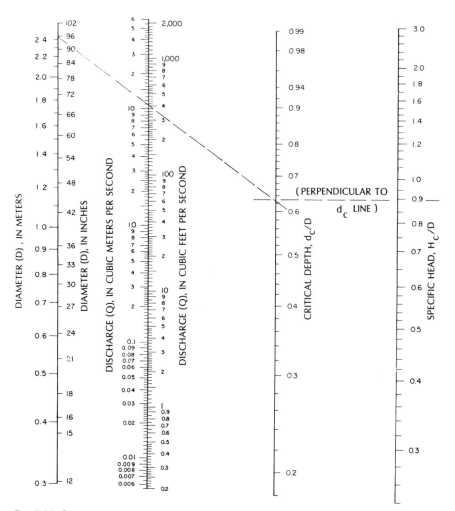

Fig. 5-23. Critical depth of flow and specific head in pipes.

b. Hydraulic Jump

Consideration is given to hydraulic jump in several aspects of design. In some cases, the purpose is to determine if a jump will occur and how it can be avoided. If the jump is necessary, the two conjugate depths are calculated as part of the computation of water-surface profiles. The general principles of these computations have already been presented in Section C above. The details of the calculations for various situations are discussed extensively in standard texts (2,3,11).

c. Appurtenances and Controls

Many of the appurtenances and controls that affect flow in sanitary sewers are discussed in Chapter 7. They include manholes, inverted siphons, bends, drop structures, flap or backwater gates, and devices for flow measurement. In addition to these, transitions, changes in conduit size, and inlets or outlets at tanks and lift stations may also serve as controls.

HYDRAULICS OF SEWERS

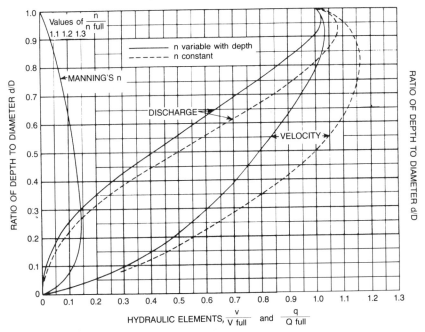

Fig. 5-24. Hydraulic properties of circular sewers.

Hydraulically, a control has two effects. First, there is a local effect, which is a significant change in flow at the control. Within a short distance there is a loss of energy and there may be a rapid change in water-surface elevation. Second, there are the upstream and downstream effects. Upstream, the control may cause a backwater or drawdown curve. Downstream, it may cause a hydraulic jump. Many of the flow situations described in Subsection C2 are the effects of controls.

Frequently the energy loss from a local effect is treated as a minor loss, and uniform flow is assumed on either side. Thus the upstream and downstream effects are not considered. Eqs. 5.19 and 5.20 are used for the losses, and the problem is reduced to finding the suitable loss or discharge coefficient for the control under consideration. Values of coefficients are found in a variety of sources, from standard texts (2,3,6,33,36) for such items as transitions and entrances, to manufacturers technical data sheets for items such as flap gates. In references they may be listed under form losses.

Energy losses are used to determine the energy grade line. On each side of the local effect, the water-surface elevations are determined by subtracting the velocity head from the energy grade line. The inverts on either side of the control can then be adjusted as necessary.

In certain cases, some engineers calculate the local effect in more detail by using the energy and momentum equations to analyze the flow in the control. Such cases include large transitions, junctions, and inverted siphons. The analysis of flow in transitions and inverted siphons is available in standard texts (2,3,6,33). A detailed and comprehensive analysis of flow in junctions based on the momentum equation has been developed, tested, and applied (12) but is not yet available in standard texts.

4. Detailed Analyses

a. Water-Surface Profiles

In sewer design there are cases in which it is inadequate to assume uniform flow or to determine elevations only at important locations. For these cases, it is necessary to calculate the actual water-surface profile. The two most common profiles are the backwater and the drawdown curves (M1 and M2). Also important are the profiles associated with the location of a hydraulic jump.

The calculation of water-surface profiles is discussed extensively in standard texts (2,3,6). Because the classical calculations are tedious step-methods requiring trial-and-error procedures at each step, they are employed infrequently. Some elaborate computer programs for water-surface profiles (37) are available, and simpler programs are available for microcomputers and programmable calculators.

b. Flow Routing and Wave Calculations

Calculation of unsteady flow is not currently a standard part of design computations for sanitary sewers. However, planning and design studies are requiring new data from flow measurement more frequently, and the interpretation of such measurements often involves unsteady flow. Flow measurements are made for determining quantities of wastewater, for surveys and monitoring of industrial discharges, for studies of infiltration and inflow, and for design of flow equalization basins at wastewater treatment plants.

The theory and analysis for routing unsteady flow and calculating wave motion have become more widespread in hydraulics texts (2,3,6,7,8), but the computations are usually too time consuming to be done by hand. Increased application will depend on the availability of computers and suitable computer programs. For certain cases, some available program packages for analysis of watersheds, floods, or combined sewers, contain programs that can be applied to sanitary sewers (38,39,40).

F. REFERENCES

1. *Nomenclature for Hydraulics,* Manual of Eng. Practice No. 43, Amer. Soc. Civil Engr., New York, 1962.
2. Chow, V.T., *Open Channel Hydraulics,* McGraw-Hill Book Co., Inc., New York, 1959.
3. Brater, E.F., and King, H.W., *Handbook of Hydraulics,* 6th Edition, McGraw-Hill, New York, 1976.
4. Rouse, H., *Elementary Mechanics of Fluids,* John Wiley & Sons, Inc., New York, 1946.
5. Rouse, H., "Fundamental Principles of Flow", in *Engineering Hydraulics,* (H. Rouse Ed.), John Wiley and Sons, New York, 1950.
6. Henderson, F.M., *Open Channel Flow,* Macmillan Company, New York, 1966.
7. Wylie, E.B. and Streeter, V.L., *Fluid Transients,* McGraw-Hill, New York, 1978.
8. Gilcrest, B.R., "Flood Routing", in *Engineering Hydraulics,* (H. Rouse Ed.), John Wiley and Sons, New York, 1950.
9. Parmakian, J., *Water Hammer Analysis,* Prentice-Hall, Inc., New York, 1961.
10. Rich, G.R., *Hydraulic Transients,* McGraw-Hill Book Co., Inc., New York, 1951.
11. Ippen, A.T., "Channel Transitions and Controls", in *Engineering Hydraulics,* (H. Rouse Ed.) John Wiley and Sons, New York, 1950.
12. Pardee, L.A., "Hydraulics of Junctions", *Office Standard No. 115,* Storm Drain Design Division, Bureau of Engineering, City of Los Angeles, Los Angeles, CA, 1968.
13. "Friction Factors in Open Channels," Prog. Rept., Task Force on Friction Factors in Open Channels, Jour. Hyd. Div., Proc. Amer. Soc. Civil Engr., Vol. 89, HY4, 1963.

14. Rouse, H., "Critical Analysis of Open-Channel Resistance." *Jour. Hydr. Div.*, Proc. Amer. Soc. Civil Engr. Vol. 91, HY4, 1965.
15. *Design Charts for Open-Channel Flow*, Hydraulic Design Series No. 3, Bureau of Pub. Roads, U.S. Govt. Printing Office, Washington, D.C., 1961.
16. *Handbook of Channel Design for Soil and Water Conservation*, Soil Conservation Services Publ. No. SCS-TP-61, U.S. Govt. Printing Office, Washington, D.C., 1954.
17. Posey, C.J., "Gradually Varied Channel Flow", in *Engineering Hydraulics*, (H. Rouse Ed.), John Wiley and Sons, New York, 1950.
18. Croley, T.E., *Hydrologic and Hydraulic Computations for Small Programmable Calculators*, Iowa Institute of Hydraulic Research, The University of Iowa, Iowa City, Iowa, 1977.
19. Williams, G.S., and Hazen, A., *Hydraulic Tables*, 3rd Ed., John Wiley & Sons, Inc., New York, 1945.
20. *Hydraulic Design Criteria*, U.S. Army Engr., Waterways Exp. Sta., Vicksburg, Miss., 1964.
21. Moody, L.F., "Friction Factors for Pipe Flow," *Trans. Amer. Soc. Mech. Engr.*, Vol. 66, pg. 671, 1944.
22. Colebrook, C.F., "Turbulent Flow in Pipes, with Particular Reference to the Transition Region between the Smooth and Rough Pipe Laws," *Journal of Institution of Civil Engineers*, 1939.
23. Wilcox, E.R., "A Comparative Test of the Flow of Water in 8-Inch Concrete and Vitrified Clay Sewer Pipe." Univ. of Washington, Exp. Sta. Series Bull. 27, 1924.
24. Yarnell, D.L., and Woodward, S.M., "The Flow of Water in Drain Tile." Dept. of Agriculture Bull. No. 854, U.S. Govt. Printing Office, Washington, D.C., 1920.
25. Camp, T.R., "Design of Sewers to Facilitate Flow." *Sew. Works Jour.*, Vol. 18, pg. 3, 1946.
26. Camp. T.R., "Discussion — Determination of Kutter's n for Sewers Partly Filled," *Trans. Amer. Soc. Civil Engr.*, Vol. 109, pg. 240, 1940.
27. Schmidt, O.J., "Measurement of Mannings Coefficient," *Sewage and Industrial Wastes*, Vol. 31, pg. 995, 1959.
28. Bloodgood, D.E., and Bell, J.M., "Manning's Coefficient Calculated from Test Data," *Jour. Water Poll. Control Fed.*, Vol. 33, pg. 176, 1961.
29. Pomeroy, R.D., "Flow Velocities in Small Sewers," *Jour. Water Poll. Control Fed.*, Vol. 39, pg 1525, 1967.
30. Neale, L.C., and Price, R.E., "Flow Characteristics of PVC Sewer Pipe," *Jour. San. Eng. Div.*, Proc. Amer. Soc. Civil Engr., Vol. 90, SA3, pg. 109, 1964.
31. Fair, G.M., Geyer, J.C., and Okun, D.A., *Water Supply and Wastewater Removal*, John Wiley & Sons, Inc., New York, 1966.
32. Babbit, H.E., and Baumann, E.R., *Sewerage and Sewage Treatment*, 8th Ed., McGraw-Hill Book Co., Inc., New York, 1958.
33. Metcalf and Eddy, Inc., *Wastewater Engineering: Collection and Pumping of Wastewater*, McGraw-Hill, New York, 1981.
34. Shields, A., *Anwendung der Aehnlichkeitsmechanik und der Turbulenz Forschung auf die Geschiebebewegung*, Mitteilungen der Preussischen Versuchsanstalt fur Wasserbau und Schiffbau, Heft 26, Berlin, 1936.
35. "Minimum Velocities for Sewers." Final Rept. Comm. to Study Limiting Velocities of Flow in Sewers, *Jour. Boston Soc. Civil Engr.*, Vol. 29, pg. 286, 1942.
36. Bauer, W.T., Louie, D.S. and Voorduin, W.L., "Basic Hydraulics", in *Handbook of Applied Hydraulics*, (C.V. Davis and K.E. Sorensen Ed.) 3rd Edition, McGraw-Hill, New York, 1969.
37. "HEC-2 Water Surface Profiles", *Generalized Computer Program 723-X6-1202A*, User's Manual, Hydrologic Engineering Center, Corps of Engineers, U.S. Army, Davis, CA, 1979.
38. Metcalf and Eddy, Inc., "Storm Water Management Model", U.S. Environmental Protection Agency, Water Quality Reports Nos. 11024 D0C07/71 through 11024 D0C10/71, 4 vols., 1971.
39. "HEC-1 Flood Hydrograph Package", *Generalized Computer Program 723-X6-L2010*, User's Manual, Hydrologic Engineering Center, Corps of Engineers, U.S. Army, Davis, CA, 1973.
40. "Computer Program for Project Formulation — Hydrology", *Technical Release No. 20*, Soil Conservation Service, U.S. Dept. of Agriculture, Washington, D.C., 1965.

CHAPTER 6

DESIGN OF SANITARY SEWER SYSTEMS

A. INTRODUCTION

Sanitary sewers transporting wastewater from one location to another must be installed deep enough below the ground surface to receive flows from the tributary area. The pipe material used must exhibit characteristics of resistance to corrosion and structural strength sufficient to carry earth backfill load and any impact and live loads. The size and slope or gradient of the sanitary sewer must be adequate for the flow rate to be carried with a velocity sufficient to inhibit deposition of solids, thereby being self-cleansing. The type of sewer pipe selected should be made based on consideration of the characteristics of the wastewater being transported and installation conditions. Manholes, junction chambers, and other structures should minimize turbulence and head loss and prevent deposition of solids. Anticipated service life, economy of maintenance, safety to personnel and the public, and convenience for connection during the useful life of the sewer must be considered.

The objective of design is to provide a sewerage system at the lowest annual cost compatible with its function and durability for the design period. The degree of complexity of the system is usually reflected in higher construction costs.

B. ENERGY CONCEPTS OF SEWER SYSTEMS

A sanitary sewer system provides a means of transportation for wastewater which utilizes the potential energy resulting from the difference in elevation of its upstream and downstream ends. Energy losses due to free falls, sharp bends, or turbulent junctions must be held to a minimum if the sewer is to operate properly.

Generally, the total available energy is utilized to maintain proper flow velocities in the sewers with minimum hydraulic head loss. However, where excess elevation differences exist, it may become necessary to dissipate excess potential energy.

Thus, sanitary sewer system design is limited by hydraulic losses which must be kept within the limits of available energy, and by utilizing available energy to maintain self-cleansing velocities. The wider the variation in rate of flow, the more difficult it becomes to meet both conditions.

Where the differences in elevation are insufficient to permit gravity flow, external energy must be added to the system by pumps. The consequences of pumping station failure due to mechanical or electrical breakdown must be considered. If a pumping station outage would result in pollution of a stream or in any way affect the health and safety of a community, the higher cost of a gravity system may be justified.

C. COMBINED VS. SEPARATE SEWERS

Many existing sewer systems collect both wastewater and stormwater. In such systems, stormwater, up to a design limit, is channeled with sanitary wastewater to the treatment plant. When combined flows exceed the sewer's capacity overflows occur and wastewater, along with stormwater, is discharged to receiving waters. When stormwater is transported with wastewater to the treatment plant, pumping and treatment costs are increased and treatment problems may occur.

Studies by the U.S. Public Health Service and Environmental Protection Agency (8,9,10) show that combined stormwater and wastewater overflows introduce large quantities of polluting materials into the nation's waters. Overflow of wastewater with stormwater may preclude the use of downstream waters for water supply. Organic matter in the wastewater may be sufficient to cause sludge deposits and depletion of dissolved oxygen in the receiving stream, and otherwise reduce the quality of water.

The only effective way to keep sanitary wastewater and storm flows apart is by means of separate systems. Combined systems are not considered good modern practice.

Combined sewers are less costly to construct than separate sanitary and storm sewers, but initial savings may prove to be false economy because of the additional cost of treatment.

D. LAYOUT OF SYSTEM

The design engineer begins a sanitary sewer system layout by selecting an outlet, determining the tributary area, locating trunk and main sewers, and determining the need for and location of pumping stations and force mains (7).

Preliminary layouts can be made largely from topographic maps and other pertinent information. In general, sanitary sewers will have a gradient in the same direction as street or ground surfaces and will be connected by trunk or main sewers.

The outlet is located according to the circumstances of the particular project. Thus, a system may discharge to a treatment plant, a pumping station, or a trunk or main sanitary sewer.

Drainage district boundaries usually conform to watershed or drainage basin areas. It is desirable to have district boundaries follow property lines so that any single lot or property is tributary to a single system. This is particularly important when assessments for sewer connections are made against property served. The boundaries or subdistricts within any assessment district may be fixed on the basis of topography, economy of sanitary sewer layout, or other practical considerations.

Trunk, main and interceptor sanitary sewers are located at the lower elevations in a given area. Some considerations which may affect the exact location are traffic conditions, type of pavement encountered, and availability of rights of way.

Consideration should be given to future needs. A system or part of one should be designed to serve not only the present tributary area but should be compatible with an overall plan to serve an entire drainage area unless this is impractical for economic reasons.

When sanitary sewers are located close to public water supplies, it is common practice to use pressure-type sewer pipe, concrete encasement of the sewer pipe or sewer pipe with joints which meet stringent infiltration/exfiltration requirements. Most building codes prohibit sanitary sewer installation in the same trench with water mains. The design engineer should check health regulations for requirements.

Manholes are located at the junctions of sanitary sewers and at changes in grade or alignment except in curved alignments (see Section E below). Also, manholes should be placed at locations that provide ready access to the sewer for preventive maintenance and emergency service. On the other hand, manholes should not be located in any low area, such as a swale gutter, where there will be a concentrated flow of water over the top that could cause excessive inflow. This may require a few additional manholes to be constructed, but it will enable the operating agency to provide better service during the life of the sewer. Inaccessible manholes are of little or no value to the operating agency.

Street intersections are common locations for manholes. When a manhole is not necessary for a present or future junction, it is better placed outside of a street intersection, but in the street right-of-way, to avoid substructure and surface congestion and to provide room for maintenance personnel.

A terminal manhole at the upper end of a sanitary sewer should be placed in the street right-of-way so that the sewer is accessible for maintenance and emergency service. It should not be located inside of the property line of the last property served. Thus for proper location of the terminal manhole the sewer should be extended to the street right-of-way if necessary.

Sanitary sewer manholes should not be located or constructed in a way that allows surface water to enter. When this is not possible, watertight manhole covers should be specified. Manholes not in the pavement, especially in open country, should have their rims set above grade to avoid the inflow of stormwater and to simplify field location.

Manhole spacing varies, reflecting available sanitary sewer maintenance methods. Typical spacings are often in the range of 90 to 150 m (300 to 500 ft). For large diameter sewers, spacing of 150 to 300 m (500 to 1,000 ft) may be used.

Tees or wyes generally should not be provided on new sewers when they are constructed. Frequently, the location of these fittings as shown on official records is incorrect, and this may cause disputes over payment for work done, such as excavations for sewer connections, based on these incorrect tee or wye locations. Another type of problem arises where the location of the tee or wye may be as indicated on the records but may not be at the location desired by the building contractor. In other cases the number of houses built may not correspond with the number of lots, in which case there may be too few or too many connections and, therefore, probably incorrectly located. Each tee or wye is a potential source of structural failure and a potential source where extraneous water may enter the sewer whether or not there is a structural failure.

When a sewer is constructed under a proposed street, wye connections should be made with a stub extended beyond the curb or edge of pavement line. If the sewer is located on one side of the street, wyes should be installed and a stub extended across the street to serve property on the other side to avoid tearing up the pavement when the building service line is constructed.

Connections to sanitary sewers should be made only with an experienced crew using equipment specifically designed for tapping a sewer. Some cities provide this service with their own staff and charge for each connection made. The authority to make these connections should be carefully controlled so that only authorized, experienced personnel make the connections.

The most common location of a sanitary sewer is at or near the center of a street or alley. A single sanitary sewer then serves both sides of the street with approximately the same length of house connection sewer. In an exceptionally wide street it may be more economical to install a sanitary sewer on each side. In such a case, the sanitary sewer may be located outside the curb between curb and sidewalk, or under the sidewalk. Normally, sidewalk locations are used only where other locations are not possible. Locating sanitary sewers within the gutter area is least desirable because of the possibility of stormwater inflow through manhole covers.

Sometimes a sanitary sewer must be located in an easement or right-of-way, for example at back property lines, to serve parallel rows of houses and residential developments without alleys. Easements must be of sufficient width to allow access for maintenance equipment, and agreements must provide the right of access for construction, inspection, maintenance and repair.

Local ordinances should provide that all above ground obstructions belonging to other utilities, such as utility poles, gas meters, and telephone junction boxes, be located as close as possible to one edge of the easement, not in the center. They should also discourage or prohibit, insofar as possible, planting of trees and shrubs, the construction of fences or retaining walls or any other above gound obstruction that would interfere with access to the entire length of the line. The ordinances should indicate where these private obstructions may be located within the easement and that they are placed there at the risk and expense of the property owner. Replacement of the obstructions that are removed to permit access should not be the responsibility of the sewer utility.

Despite such provisions access sometimes becomes difficult or impossible. It becomes irritating to both the property owners and the utility. Sewer locations in streets or other public properties, therefore, are greatly preferred.

E. CURVED SANITARY SEWERS

The design and installation of large-diameter sanitary sewers laid on curves is a generally accepted practice in sanitary sewer design (1). Several benefits are derived from this type of installation: The installation of curved sanitary sewers will result in economies over straight-run sewers by eliminating manholes needed at each abrupt change of direction; the installation of sanitary sewers parallel to or on the centerline of a curved street makes it easier to avoid other utilities; such installations will allow manhole locations away from street intersections. The design of curved main or trunk sanitary sewers allows the engineer to follow topographic contours for the desired alignment and simplifies the maintenance of a uniform gradient. Inspection and maintenance requirements generally determine minimum diameters of curved sewers.

1. Rigid Pipe

The installation of rigid sewer pipe on a curve is accomplished by deflecting the pipe joint from the normal straight position. The maximum permissible deflection must be limited so that satisfactory pipe joint performance is not affected. When rigid sewer pipe is installed on a curve, it is advisable that the manufacturer of the sewer pipe be consulted and the maximum allowable pipe joint deflection determined.

The radius of curvature which may be obtained by deflection of straight sewer pipe is a function of the deflection angle per pipe joint (joint openings),

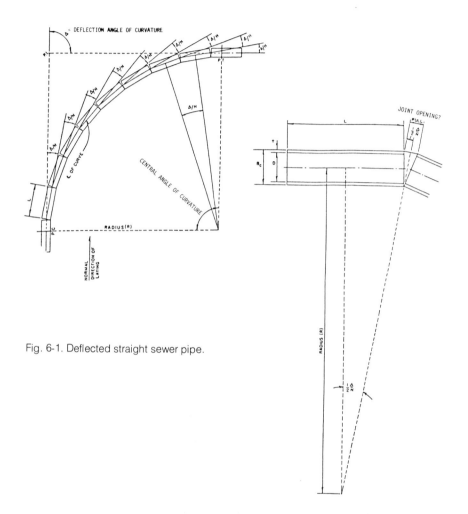

Fig. 6-1. Deflected straight sewer pipe.

Fig. 6-2. Curved alignment using deflected straight sewer pipe.

the diameter of the sewer pipe and the length of the sewer pipe sections. The radius of curvature (R) is computed by the equation:

$$R = \frac{L}{2\left[\tan \frac{1}{2}\left(\frac{\Delta}{N}\right)\right]} \tag{6.1}$$

where R = radius of curvature, in meters (feet); L = length of sewer pipe sections measured along centerline, in meters (feet); Δ = total central (or deflection) angle of the curve, in degrees (radians); N = number of sewer pipe sections; and Δ/N = total central or deflection angle for each sewer pipe, in degrees (radians).

From Fig. 6-1, the angular deflection from the tangent of the circle, 1/2 (Δ/N), is further defined as:

$$\frac{1}{2}\left(\frac{\Delta}{N}\right) = \sin^{-1}\left(\frac{Deflection}{2B_c}\right) \tag{6.2}$$

where *Deflection* = joint opening, in meters (feet); and B_c = outside sewer pipe diameter, in meters (feet).

Fig. 6-2 shows sewer pipe installed on a curved alignment using straight sewer pipe. As an alternate to deflecting the straight sewer pipe, the desired radius of curvature may be based on the fabrication of radius sewer pipe. Radius sewer pipe, also referred to as bevelled or mitered pipe, incorporates the deflection angle into the sewer pipe joint. The sewer pipe is manufactured with one side of the pipe shortened; the amount of shortening or drop is dependent on manufacturing feasibility. With the possibility of greater deflection angles per joint, shorter radii can be obtained with radius sewer pipe than with deflected straight sewer pipe.

The radius of curvature which may be obtained with radius sewer pipe is a function of the deflection angle per joint, diameter of the sewer pipe, length of sewer pipe section and wall thickness. It is computed from the equation:

$$R = \frac{L}{\tan\left(\dfrac{\Delta}{N}\right)} - \left(\frac{D}{2} + t\right) \tag{6.3}$$

where R = radius of curvature; Δ = total central or deflection angle of curve, in degrees (radians); N = number of radius sewer pipe sections; L = standard sewer pipe length being used, in meters (feet); D = inside diameter of sewer pipe, in meters (feet); and t = wall thickness of the sewer pipe, in meters (feet).

From Fig. 6-3, the radius of curvature (R) can be expressed in terms of the bevel and is given by the equation:

$$R = \frac{L(D + 2t)}{Drop} - \left(\frac{D}{2} + t\right), \text{ or } \left(\frac{L}{Drop} - \frac{1}{2}\right) B_c \tag{6.4}$$

where B_c = outside diameter of the pipe, in meters (feet); and *Drop* = length pipe is shortened on one side.

It is essential to coordinate the design of curved sanitary sewers with the radius pipe manufacturer. Fig. 6-4 shows sewer pipe installed on a curved

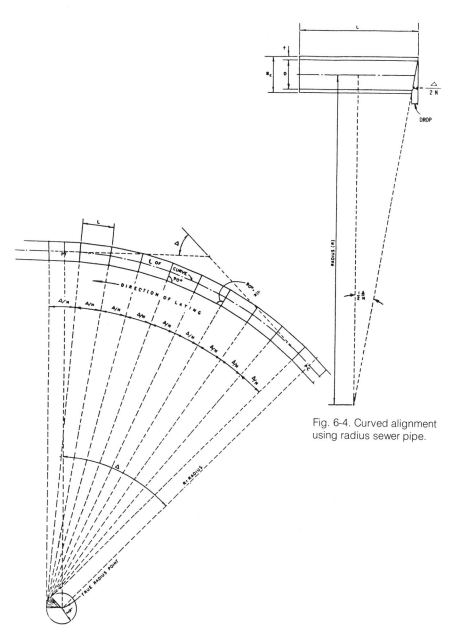

Fig. 6-4. Curved alignment using radius sewer pipe.

Fig. 6-3. Radius sewer pipe.

alignment using radius pipe.

The minimum radius of curvature obtained from Eqs. 6.1 and 6.3 is within a range of accuracy that will enable the sewer pipe to be readily installed on the required alignment.

2. Flexible Pipe

The installation of flexible sewer pipe on a curve is accomplished by controlled longitudinal bending of the pipe and deflection of the pipe joint. When flexible sewer pipe is installed on a curve, it is advisable that the manufacturer of the pipe be consulted on the minimum radius of bending and axial joint deflection.

Permissible joint deflection may be significant when gasketed joints, which are designed for deflection, are provided on thermoplastic pipe. Solvent cement or fusion joints provide no flexibility.

Mathematical relationships for longitudinal bending of pressurized tubes have been derived (11,12). One critical limit to bending of flexible pipe is long-term flexural stress. However, axial bending causes a very small amount of ovalization or diametric deflection of the pipe.

The equation for allowable bending stress (S_b) in pascals (pounds per square foot) is:

$$S_b = (HDB)\frac{T}{F} \tag{6.5}$$

where HDB = hydrostatic design basis of pipe, in pascals (pounds per square foot); F = safety factor (2.0 is suggested for non-pressure pipe); and T = temperature rating factor [1.0 at 73.4°F (23°C)].

Also to be considered is the mathematical relationship between stress and moment induced by longitudinal bending of pipes:

$$M = \frac{S_b I}{c} \tag{6.6}$$

where M = bending moment, in newton-meters (foot-pounds); S_b = allowable bending stress, in pascals (pounds per square foot); c = distance from extreme fiber to neutral axis = OD (outside diameter)/2, in meters (feet); and I = moment of inertia, in meters (feet) to fourth power.

With reference to Fig. 6-5, and assuming that the bent length of pipe conforms to a circular arc after backfilling and installation, the minimum radius of the bending circle (R_b) can be found by Timoshenko's equation (13):

$$R_b = \frac{EI}{M} \tag{6.7}$$

where E = Young's modulus of elasticity, in pascals (pounds per square foot).
Combining Eqs. 6.6 and 6.7 gives:

$$R_b = \frac{E}{2S_b}(OD) \tag{6.8}$$

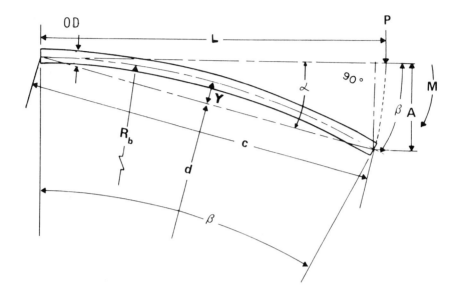

Fig. 6-5. Flexible pipe allowable blend.

The central angle β subtended by the length of pipe is:

$$\beta = \frac{360}{2\pi R_b}(L) = \frac{57.30}{R_b}(L) \qquad (6.9)$$

where L and R_b are both in the same units, and the angle of lateral deflection α [in degrees (radians)] of the curved pipe from a tangent to the circle is:

$$\alpha = \frac{\beta}{2} \qquad (6.10)$$

with L = pipe length, in meters (feet). The offset at the end of the pipe from the tangent to the circle, A, in meters (feet), is:

$$A = 2R_b \sin\left(\frac{\beta}{2}\right)^2 = 2R_b (\sin \alpha)^2 \qquad (6.11)$$

Because of the characteristics of a particular joint design, it is possible that a manufacturer's recommended bending radius may be greater or lesser than those calculated.

When change of direction in a flexible pipeline exceeds the permissible bending deflection angle for a given length of pipe, the longitudinal bending required should be distributed through a number of pipe lengths, or an elbow should be used.

F. TYPE OF CONDUIT

A variety of materials are available to the design engineer for use in sanitary sewers. (A discussion of properties and applications of the different sewer pipe materials is presented in Chapter 8.) The type of materials a design engineer chooses to specify is dictated by several factors along with sound engineering judgment. These factors include the type of wastewater to be transported (residential, industrial, or a combination of both), the type of soil and trench load conditions, and the bedding and initial backfill material available. (Information on structural design for both rigid and flexible conduits is presented in Chapter 9.)

G. VENTILATION

Forced ventilation of a sanitary sewer is generally considered a special application used to solve a specific problem. Natural ventilation from manholes, building vents and flow variations of the wastewater, in most cases, is adequate to provide oxygen in the sanitary sewer atmosphere.

When forced ventilation is required, special airtight or pressure manhole covers must be used and the air exhausted to a high stack or a deodorizing process.

The design requirements for forced ventilation are specific to a particular sanitary sewer layout and design. They should not be interpreted as minimizing in any way the need for gas detection and ventilation before and during maintenance operations to eliminate any possible danger to sanitary sewer workers.

One of the largest ventilation stations in use to date is the Kain Station in Sydney, Australia, which exhausts 400 m^3/min (14,000 cfm) by way of a stack 40 m (130 ft) high. The air is treated for odor reduction, yet at times there have been odor complaints from residents in the vicinity.

H. DEPTH OF SANITARY SEWER

Sanitary sewers should be installed at such depths that they can receive the contributed flows from the tributary area by gravity flow. Deep basements and buildings on land substantially below street level may require individual pumping facilities. Sufficient sanitary sewer depth must be provided to prevent freezing and backflow of wastewater through connections.

No single method prevails for determining the minimum depth for a sanitary sewer. One suggestion is that the top of the sanitary sewer should not be less than 1 m (3 ft) below the basement floor of the building to be served. Another rule places the invert of the sanitary sewer not less than 1.8 m (6 ft) below the top of the house foundation. The latter assumes that it is not necessary for a sanitary sewer to serve basement drains, which results in a considerable saving by reducing sanitary sewer depth. It also has the advantage of preventing the connection of exterior basement wall footing drains to the sanitary sewer. This, however, is acceptable only where few basements have sanitary facilities or if basement sump pumps are utilized.

It is common to lay house connections at a slope of 2%, with a minimum slope of 1%. In some developments in which houses are set well back from the street, the length and slope of the house connection may determine minimum sewer depths. High and unjustified costs may preclude the lowering of a whole sewer system to provide service for only a few houses.

Where houses have no basements, sewers may be built at shallower depths. In business or commercial districts, however, it may be necessary to lay sewers as deep as 3.6 m (12 ft) or more to accommodate the underground facilities in such areas.

As sewers usually are laid in public streets or easements, consideration must be given in design to the prevention of undue interference with other underground structures and utilities. The depth of the sanitary sewer is usually located such that it can pass under all other utilities with the possible exception of storm sewers.

When sewers are to be laid at shallow depths, consideration should be given to live and impact loads since special requirements may be necessary in the selection and installation of the pipe. (Structural requirements are detailed in Chapter 9.)

I. FLOW VELOCITIES AND DESIGN DEPTHS OF FLOW

A sanitary sewer has two main functions: To convey the designed peak discharge and to transport solids so that deposits are kept to a minimum. It is essential, therefore, that the sanitary sewer have adequate capacity for the peak flow and that it function at minimum flows without excessive maintenance and generation of odors.

It is customary to design sanitary sewers with some reserve capacity. Generally, sanitary sewers through 375 mm (15 in.) in diameter are designed to flow half full. Larger sanitary sewers are designed to flow three-fourths full.

Velocity determination is based on calculated peak flow, which is commonly considered to be twice the average daily flow. Some design engineers use peaking factor curves to determine the peak-to-average ratio in designing for peak conditions. This ratio can vary from 1.5 to 4.0 depending on the tributary area and population. The average daily flow is the average 24-hr discharge calculated during a 1-yr period.

Accepted standards dictate that the minimum design velocity should not be less than 0.60 m/sec (2 fps) or generally greater than 3.5 m/sec (10 fps) at peak flow. A velocity in excess of 3.5 m/sec (10 fps) can be tolerated with proper consideration of pipe material, abrasive characteristics of the wastewater, turbulence, and thrust at changes of direction. The minimum velocity requirement is necessary to prevent the deposition of solids.

Special attention must be given to the early years that the system is used, as initial flows may be substantially lower than design flows and the velocities well below minimum. The design engineer may elect to use a greater slope or the owner may desire to inititate a well planned line cleaning maintenance program.

(Information on flow and velocities as they relate to hydrogen sulfide generation is presented in Chapter 4.)

J. INFILTRATION/INFLOW

In many sanitary sewers, extraneous flow consisting of infiltration and inflow is a major cause of hydraulic overloading of both the collection system and treatment plant (2,3,4). To handle this excess flow it may become necessary to construct relief sanitary sewers and expand existing treatment facilities. These measures add to the cost of the system. Other expenses incurred because of this unwanted flow include: Higher pumping cost; replacement cost of failed sanitary sewers and surface pavements resulting from soil flushing into the sewer; and higher maintenance cost resulting from soil deposits in sanitary sewers and root penetration into leaky joints.

Infiltration and inflow can contribute substantially to sewer flows. Inflow is often the result of deliberately planned or expediently devised connections of storm water sources into sanitary sewer systems, or from unintentional storm water sources that result from structure design, location or deterioration of sanitary sewers. The inflow from such sources can be prevented or corrected by regulation and inspection procedures aimed at enforcing regulations relating to sanitary sewer connections.

Infiltration results from the age of the structure, soil conditions, materials and methods of construction. It must be taken into consideration in the design, construction, and inspection of sanitary sewer systems. (A more detailed discussion of infiltration and related matters is found in Chapter 3.)

K. INFILTRATION/EXFILTRATION AND LOW-PRESSURE AIR TESTING

1. Infiltration-Exfiltration Test Allowance

The most effective way to control infiltration, and at the same time to assure structural quality and proper installation of the new sanitary sewer, is to establish and enforce a maximum leakage limit as a condition of job acceptance. Limits may be stated in terms of water leakage quantities and should include both a maximum allowable test section rate and a maximum allowable system average rate. Current information indicates that a maximum allowable infiltration rate of 50 to 200 gal/in. diam/mi (5 to 19 L/mm diam/km) of sewer pipe per day can be achieved without additional construction costs.

Manholes may be tested separately and independently. An allowance for manholes of 0.1 gal/hr/ft diam/ft head (4L/hr/m diam/m head) would be appropriate.

2. Infiltration-Exfiltration Testing

When groundwater is observed to be at least 1.2 m (4 ft) above the top of the sewer pipe, the infiltration test will determine the integrity of the sewer line. Any leakage can be measured with a V-notch weir or similar flow measuring device. If no leakage is observed, it can be assumed that the line passes the test.

If the groundwater level is not at least 1.2 m (4 ft) above the top of the sewer pipe, then an exfiltration test is required. This is performed by plugging the manhole at the lower end of the test section and filling the line with water. If leakage does not exceed the limits specified, then the section tested is accepted. If leakage exceeds the limits specified, the leak must be located and repaired.

3. Low-Pressure Air Testing

A low-pressure air test may also be used to detect leaks in sewer pipe where hydrostatic testing is not practical (5,6). Because of the physical difference between air and water and the difference in behavior under pressure conditions, the air test cannot be related to the water test, although either test can be used with confidence to prove the integrity of the sewer line. The air test depends on porosity, moisture content, and wall thickness of the sewer pipe. A well constructed sanitary sewer which is impervious to water may still have some air loss through the sewer pipe wall. In applying low-pressure air testing to sanitary sewers designed to carry fluid under gravity conditions, it is necessary to distinguish between air losses inherent in the type of sewer pipe material used and those caused by damaged or defective pipe joints.

Two air test methods used are the *constant pressure* method and the *time pressure* method, with the latter the most commonly used. The constant pressure method utilizes an air flow measuring device operated at 3 psi (20 k Pa) greater pressure than the average back pressure of any groundwater. In the time pressure drop method, the air supply is disconnected and the time required for the pressure to drop from $3\frac{1}{2}$ psi (24kPa) to $2\frac{1}{2}$ psi (17 kPa) gauge is determined. Test procedures and calculations are available from ASTM. Ramseier's work (5,6) also should be considered in air testing.

In applying the low-pressure air test, the following factors should be understood and precautions followed during the test.
- The air test is intended to detect defects in the sewer line and establish the integrity of the line under sewer conditions.
- Since the pipe will be in a moist environment when in service, testing the pipe in wet conditions is appropriate.
- Plugs should be securely braced.
- Plugs should not be removed until all air pressure in the test section has been reduced to ambient pressure.
- For safety reasons, no one should be allowed in the trench or manhole while the test is being conducted.
- The testing apparatus should be equipped with a pressure relief device to prevent the possibility of loading the test section with full compressor capacity.

Pipe that is large enough to permit personnel to conduct interior inspections can be accepted on the basis of such inspection, plus air testing of individual joints if required. When individual joints are tested, allowable leakage is usually the computed rate per foot (meter) of pipe times the distance between joints. In practice, however, it is a go or no-go test.

L. DESIGN FOR VARIOUS CONDITIONS

1. Open Cut

The load on a sanitary sewer built in open cut is a function of the bedding, trench width, backfill material and any superimposed loads; consideration must be given to all of these elements. Backfill material placed 0.3 to 0.6 m (12 to 24 in.) over the top of the sewer pipe should be free of rocks or stones larger than 50 mm (2 in.) in diameter to avoid damage to the sewer pipe. (Chapter 9, devoted to loads on sewer pipe, presents details of this phase of design.)

2. Tunnel

A thorough knowledge of tunnel construction methods is required before designing sanitary sewers for tunnel placement. This is especially necessary to effect economy of construction in this costly type of work. (Tunneling methods are covered in Chapter 9.)

3. Sanitary Sewers Built in Rock

Where sanitary sewers are built in rock trenches, special attention should be given to the method of bedding to avoid damage due to contact with the rock. Adequate clearances should be provided between the bottom and sides of the sanitary sewer and the adjacent rock trench. Granular bedding or a concrete cradle normally is provided. If blasting is anticipated in the area, the concrete cradle should be separated from any rock by a granular cushion.

4. Exposed Sanitary Sewers

Sometimes sanitary sewers have to be built above the ground surface. In these cases, the sewer pipe will be carried on supports or designed as a self-supporting span. In climates where freezing conditions exist, a method of protection should be employed.

5. Foundations

Knowledge of foundation conditions should be obtained by borings, soundings, or test pits along the route of a sanitary sewer prior to design. Unstable foundation soils encountered in the form of silt, peat bog, saturated sand, or other soft or flowing material require special bedding and must be considered in the design of a sanitary sewer under these conditions. If encountered during construction, costs usually will be higher than if anticipated beforehand.

Where the conditions are not severe, it may be possible to stablize the trench bottom by placing a layer of crushed rock below the sewer pipe. The rock must be fine enough, or contain fines, so that settlement will not result from unstable bottom material flowing into the voids. Concrete or wooden cradles often will suffice to spread the load in wet or moderately soft foundations. In some cases, underdrains laid beneath the sanitary sewer or well points will remove water held in the soil and permit dry construction, and they may eliminate the need for special foundations. Pipe joints that are tight, yet flexible, are particularly important when sanitary sewers are installed in areas with unstable soils.

6. Sanitary Sewers on Steep Slopes

Erosion control devices or methods may be required on steep slopes. It may also be necessary to provide anchorage or cut-off dams to prevent the sewer pipe from creeping downhill, or to prevent water from flowing along the pipe and causing the trench to wash out. If drop manholes are used and the flow is heavy, special energy dissipators may be required in the form of a special manhole bottom or a water cushion.

Table 6-1 Typical Computation Form for Design of Sanitary Sewers

Line No. (1)	Location (2)	Manhole No.		Length (ft) (5)	Area		Max Flow				Min Flow				Slope of Sewer (16)
								Sewage and Infiltration				Sewage and Infiltration			
		From (3)	To (4)		Increment (acre) (6)	Total (acre) (7)	Infiltration (mgd) (8)	Sewage (mgd) (9)	(mgd) (10)	(cfs) (11)	Infiltration (mgd) (12)	Sewage (mgd) (13)	(mgd) (14)	(cfs) (15)	

Note: Ft × 0.305 = m; acre × 0.0405 = ha; mgd × 3.785 = m³/day; cfs × 1.7 = m³/min; cfs/acre × 4.2 = m³/min/ha; in. × 2.54 = cm

SEWER SYSTEM DESIGN

Diam (in.) (17)	Capacity Full (cfs) (18)	Velocity Full (fps) (19)	Min Velocity (fps) (20)	Max Velocity (fps) (21)	Max Depth (ft) (22)	Max Velocity Head (ft) (23)	Max Energy Head (ft) (24)	Manhole Loss: Transition + Curve + Junction (ft) (25)	Manhole Invert Drop (ft) (26)	Fall in Sewer (ft) (27)	Sewer Invert Elevation		Elevation Ground Surface	
											Upper End (28)	Lower End (29)	Upper End (30)	Lower End (31)

M. RELIEF SEWERS

An overloaded existing sanitary sewer may require relief, with the relief sewer constructed parallel to the existing line to divert flows to alternate outlets.

In the design of a relief sanitary sewer, it must be decided whether (a) the proposed sewer is to share all rates of flow with the existing sanitary sewer; or (b) it is to take all flows in excess of some predetermined quantity; or (c) it is to divert a predetermined flow from the upper end of the system. The topography, available outlets, and available head may dictate which alternate is selected. If flows are to be divided according to some ratio, the inlet structure to the relief sanitary sewer must be designed to divide the flow. If it is to take all flows in excess of a predetermined quantity, the excess flow may be discharged over a side-overflow weir or through a regulator to the relief sanitary sewer. If flow is to be diverted in the upper reaches of a sewer system, the entire flow at the point of diversion may be sent to the relief sanitary sewer or the flow may be divided in a diversion structure.

An examination of flow velocities in the existing and relief sanitary sewers may determine the method of relief to use. If self-cleaning velocities cannot be maintained in either or both sanitary sewers when a division of flows is used, excess maintenance and H_2S conditions may result. If, on the other hand, the relief sanitary sewer is designed to take flows in excess of a fixed quantity, the relief sanitary sewer itself will stand idle much of the time, and deposits in it may cause similar problems. Engineering judgment is required in deciding which method of relief to use. In some cases it might be better to design the new sanitary sewer with sufficient capacity to carry the total flow and to abandon the old one.

N. ORGANIZATION OF COMPUTATIONS

The first step in the hydraulic design of a sewer system is to prepare a map showing the locations of all the required sanitary sewers and from which the area tributary to each point can be measured. Preliminary profiles of the ground surface along each line also are needed. These should show the critical elevations which will establish the sewer pipe grades, such as the basements of low-lying houses and other buildings and existing sanitary sewers which must be intercepted. Topographic maps are useful at this stage of the design.

Several trial designs may be required to determine which one will properly distribute the available hydraulic head. Time may be saved if grades are established tentatively by graphical means on profile paper before selecting final grades and computing the sewer pipe invert elevations.

Design computations, being repetitious, may best be done on tabular forms permitting both wastewater quantity and the sanitary sewer design calculations to be placed on the same form. The form shown in Table 6-1 is fairly comprehensive and can be adapted to the particular needs of the designer. It is convenient in using this form to record the data for the sewer reaches on alternate lines, reserving intervening lines for the data on transition losses and invert drops.

In using forms of this type, it is assumed that uniform flow exists in all reaches. The form, therefore, is not recommended where a detailed analysis of the wastewater surface profile is to be based on non-uniform flow.

The use of Table 6-1 for sanitary sewer design requires supplementary charts, graphs, or tables for calculating wastewater flows and hydraulic data.

Methods for computing the quantities of wastewater flow listed under Columns 8 through 15 are described in Chapter 3. Methods of calculating the hydraulic data in Columns 16 through 25 are set forth in Chapter 5. If the value of Column 26 is positive, an invert elevation drop is indicated; if negative, an invert elevation rise is indicated but would not be installed because of the wastewater effect during low flows, thus a value of zero then should be recorded in the column.

O. REFERENCES

1. "Feasibility of Curved Alignment for Residential Sanitary Sewers" Fed. Housing Admin. Rept. No. 704, U.S. Gov. Printing Office, Washington, D.C.
2. Santry, I.W., Jr., "Infiltration in Sanitary Sewers", *Journal Water Pollution Control Fed.*, Vol. 36, pg. 1256 (1964)
3. "Municipal Requirements for Sewer Infiltration", *Pub. Works*, Vol. 96, 6, pg. 158 (1965)
4. Brown, K.W., and Caldwell, D.H., "New Techniques for the Detection of Defective Sewers", *Sewage and Industrial Wastes*, Vol. 29, pg. 963 (1957)
5. Ramseier, R.E., and Rick, G.C., "Low Pressure Air Test for Sanitary Sewers", *Journal San. Eng. Div.*, Proc. Amer. Soc. Civil Engr. Vol. 90, SA2 1 (1964)
6. Ramseier, R.E., "Testing New Sewer Pipe Installations", *Jour. Water Poll. Control Fed.*, (April 1972)
7. *Sewer System Evaluation, Rehabilitation and New Construction – A Manual of Practice*, Environmental Protection Agency Publication 600/2-77-017d (Dec. 1977), National Technical Information Service, Springfield, VA 22161
8. Pomeroy, R.D., "Flow Velocities in Small Sewers", *Journal Water Pollution Control Fed.*, Vol. 39, pg. 1525 (1967)
9. "Pollution Effects of Stormwater and Overflows from Combined Sewer Systems," U.S. Public Health Service Publ, No. 1246, U.S. Govt. Printing Office, Washington, D.C. (1964)
10. "Urban Storm Runoff and Combined Sewer Overflow Pollution", U.S. Environmental Protection Agency EPA Publ. No. 11023 (Dec. 1970)
11. Reinhart, F.W. "Long-Term Working Stress of Thermoplastic Pipe", *Society of Petroleum Engineers Journal*, Vol. 17, No. 8 (August, 1961) pg. 75
12. Reissner, E., "On Finite Bending of Pressurized Tubes", *Journal of Applied Mechanics*, Transactions Amer. Soc. Mechanical Engr. (Sept. 1959), pp. 386-392
13. Timoshenko, S., *Strength of Materials*, D. Van Nostrand Co., N.Y. 1948

CHAPTER 7

APPURTENANCES AND SPECIAL STRUCTURES

A. INTRODUCTION

Certain appurtenances are essential to the proper functioning of sanitary sewer systems. They may include manholes, terminal cleanouts, service connections, inverted siphons, junction chambers and other structures or devices of special design.

Many states have established criteria through their regulatory agencies which govern, to some extent, the design and construction of appurtenances to sanitary sewer systems. In addition, each municipal engineering office or private office acting for a municipality has its own design standards. It is to be expected, therefore, that many variations will be found in the design of even the simplest structures. The discussion to follow is limited to a general description of each of the various appurtenances, with special emphasis on the features considered essential to good design.

B. MANHOLES

1. Objectives

A manhole design should pass at least these major tests. It should:
(a) Provide convenient access to the sewer for observations and maintenance operations;
(b) cause a minimum of interference with the hydraulics of the sewer;
(c) be durable and generally a watertight structure; and
(d) be strong enough to support applied loads.

2. Manhole Spacing and Location

See Section D of Chapter 6.

3. General Shape and Dimensions

Most manholes are essentially cylindrical in shape, with the inside dimensions sufficient to perform inspecting and cleaning operations without difficulty. On small sewers, a minimum inside diameter of 1.2 m (4 ft) at the bottom has been widely adopted in the U.S. A diameter of 1.0 m (3 ft) is more common in some other countries. The diameter is generally constant up to a cone at the top where the diameter is reduced to receive the frame and cover (Fig. 7-1a). In some areas where brick manholes are used, the 1.2 m (4 ft)-diam. cylinder is tall enough for adequate working space and, above that, a 1.0 m (3 ft) shaft is constructed up to the cone (Fig. 7-1c). It has become common practice in recent years to use eccentric cones, especially in precast concrete manholes, thus providing a vertical side for the steps (Fig. 7-1b). Most often the orientation places the steps over the bench, but some designs place the steps opposite the outlet pipe thus preserving the maximum working

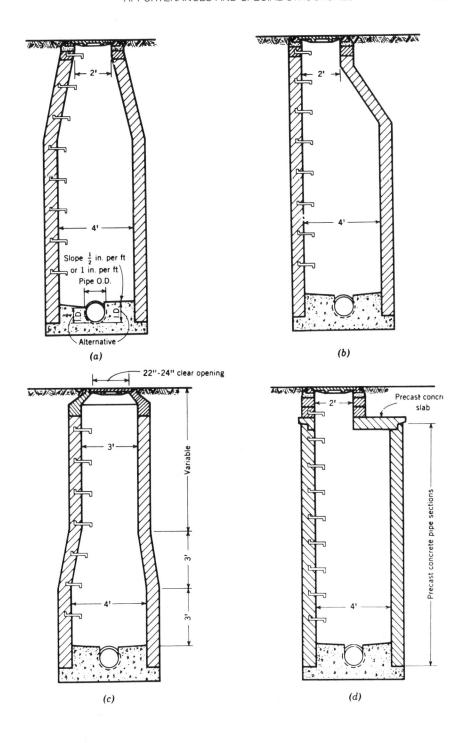

Fig. 7-1. Typical manholes for small sewers (ft × 0.3 = m; in. × 2.54 = cm).

space on the bench.

Another design used under special conditions, especially where a larger working space is needed, specifies a reinforced concrete slab instead of a cone, as shown in Fig. 7-1d. This is applicable whether the working space is circular or rectangular. The slab must be reinforced suitably to withstand traffic and earth loads.

4. Shallow Manholes

Irregular topography sometimes results in shallow manholes. A manhole of standard design does not provide a space in which a man can work effectively if the depth is only 1 to 1.5 m (3 to 4.5 ft). An extra-large cover with a 0.75 to 0.9 m (30 to 36-in.) opening helps improve this condition. A manhole that is cylindrical up to a flat slab at the surface is suitable if the head room is 1.2 m (4 ft). Usually the best option is to plan on maintenance work being done from the surface. Sometimes slots have been provided no wider than the diameter of the sewer. A special foundation is needed if a slot access of this type must support traffic loads.

5. Construction Material

The materials most commonly used for manhole walls include precast concrete sections and cast-in-place concrete. Manholes built before the middle of this century were usually made of brick, followed by concrete block with occasional cast-in-place designs. Since then, precast manhole sections have become dominant, at least in the U.S., and pre-formed fiberglass-reinforced manholes are being used in some places. The sections are available in various heights and include properly spaced steps. Precast manhole bottoms also have been used in some places.

Transition sections are furnished to reduce the diameter of the manhole at the top to accommodate the frame and cover. It is common practice to allow three or four courses of brick or concrete rings just below the rim casting to permit easier future adjustment of the top elevation.

Brick or concrete block manhole walls normally are built 200 mm (8 in.) thick at the shallower depths, and may increase to 300 mm (12 in.) below 2.5 to 3.5 m (8 to 12 ft) from the surface. Joints should be filled completely with cement mortar. The outside walls of brick or block manholes should be plastered with cement mortar not less than 13 mm (½ in.) thick. In wet areas, a bituminous damp-proofing compound is often applied to the exterior of the cement mortar.

It is difficult to make brick manholes watertight, and precast manholes may leak because of imperfect sealing of the joints. Preferably the sections of precast manholes are joined in the same manner as in a pipeline, using elastomeric gaskets or a joint filler of proven effectiveness.

6. Frame and Cover

The manhole frame and cover normally are made of cast or ductile iron. The cover is designed with these objectives:

(a) *Provision of an adequate aperture for access to the sewer.* The most common practice in the U.S. is to require a 600-mm (24-in.) clear opening. Sometimes 550-mm (22-in.) openings have been used. It is difficult for a large man, properly equipped, to enter through a 550-mm (22-in.) opening. For sewers up to 900 mm (3 ft) in diameter, covers sometimes are used that have a clear opening of 900 mm (3 ft) to accommodate cleaning equipment equalling the sewer diameter. However, collapsible cleaning equipment is the rule for large sewers. Many authorities do not use openings larger than 600 mm (2 ft); openings larger than 900 mm (3 ft) are used in special cases.

(b) *Adequate strength to support superimposed loads.* A typical traffic-weight cover for a 600-mm (24-in.) clear opening weighs about 75 kg (160 lb) and the frame about the same amount or somewhat more. Weights up to 200 kg (440 lb) are specified in some places as the total for cover and frame. Lighter weights may be used where there is no danger that they will be subjected to heavy loads.

(c) *A good fit between cover and frame,* so there will be no rattling in traffic. It is usually specified that the seat in the frame on which the cover rests and the matching face of the cover be machined to assure good fit.

(d) *Provision for opening.* Most commonly this takes the form of a notch at the side where a pick or bar can be used to pry the cover loose, often supplemented by a pick hole a short distance in from the edge.

(e) *To stop earth and gravel from falling into the sewer when the manhole must be opened.* To intercept sticks or earth inserted or falling through the pick hole, a dust pan sometimes is placed under the cover. Usually this is an iron disc, slightly smaller than the manhole opening, resting on lugs at the base of the frame. A polyethylene bowl-shaped diaphragm is now on the market that will retain dirt, and it is gasketed so that it is supposed to prevent the inflow of water to the sewer. Rubber gaskets sometimes are laid on the seat under the cover to maintain tightness in low areas subject to flooding.

(f) *Resistance to unauthorized entry.* The principal defense against a cover being lifted by children is its weight, but more persistent and competent vandals bent on throwing debris into a manhole are not deterred easily. An emergency measure in an area particularly plagued by mischief of this sort or by illegal disposals is a covering of planks over the channel. Sometimes covers are bolted in place and occasionally a lock is provided. The theft of manhole covers is a problem in some places.

7. Connection Between Manhole and Sewer

Differential settling of the manhole and the sewer sometimes breaks the sewer pipe. A pipe joint just outside the manhole permits flexibility and lessens this danger. If the soil conditions are quite unstable, a second joint within 1 m (3 ft) of the first may be necessary. To accomplish this purpose, the joints must not be rigid. Elastomeric gaskets and couplings are available to form flexible, watertight connections between the manhole and the sewer pipe, thus allowing not only flexure but also a minor amount of differential settlement that otherwise would break the pipe (Fig. 7-2).

8. Steps

Manhole steps should be wide enough for a man to place both feet on one step, with a design to prevent lateral slippage off the step, and far enough from the wall to be easy to stand on. They are generally spaced at 0.3 to 0.4m (12 to 16-in.) intervals. Attention must be given to prevailing safety regulations such as those issued by the Occupational Safety and Health Administration in the U.S.

Types that have been most commonly used are made of ductile iron or shaped from 20 to 25-mm (3/4 or 1-in.) galvanized steel or wrought iron bars. These metals corrode in the moist atmosphere prevailing in most manholes, but under normal conditions corrosion is not rapid. If the wastewater contains

Fig. 7-2. a & b. Flexible joint connection between pipe and manhole.

much hydrogen sulfide and especially if there also is much turbulence, the lower steps will fail in a few years. Steps formed of 15-mm (5/8-in.) AISI Type 304 stainless steel rods have been used in a few places. They should have a long life under most conditions. In recent years steps made of plastics or steel armored with plastic have been used. Aluminum has also been successfully used in many sanitary sewers. However, it is a very corrodible metal under conditions sometimes encountered in sewers.

In some areas, steps are omitted entirely. This is sometimes done to reduce the danger that a corroded step might break under a person's weight.

9. Channel and Bench

A channel of good hydraulic properties is an important objective that frequently is not realized because of careless construction. The channel should be, insofar as possible, a smooth continuation of the pipe. In fact, the pipe sometimes is laid through the manhole with the top half removed to provide the channel. If this is done merely by breaking out the upper half it is difficult to make a satisfactory channel. The best practice seems to be to lay a neatly cut half pipe, then build up the sides with concrete, or to use steel forms.

The completed channel cross-section should be U-shaped. Some engineers specify a channel constructed only as high as the centerline of the pipe on small sizes; others require that the height be three-fourths of the diameter or the full diameter. For sizes 375 mm (15 in.) and larger, the required channel height rarely is less than three-fourths of the diameter.

Loss of energy caused by expansion and contraction of the stream from pipe to U-shaped channel to pipe, with the pipe running full, should be less than 3% of the velocity head if the U-channel is the full depth of the pipe. It will be much greater with a channel that allows the water to swirl out over the bench. Close attention to detail is required to secure well constructed U-shaped channels.

The bench should provide good footing for a workman and a place where minor tools and equipment can be laid. It must have enough pitch to drain to the channel but not too much. A slope of 1 in 12 is common, but 1 in 24 is specified in some areas to provide a safer footing.

In the past it was ordinary practice to allow an arbitrary drop of 30 mm (0.1 ft) in the invert across the manhole, or a slope of 0.025, regardless of the

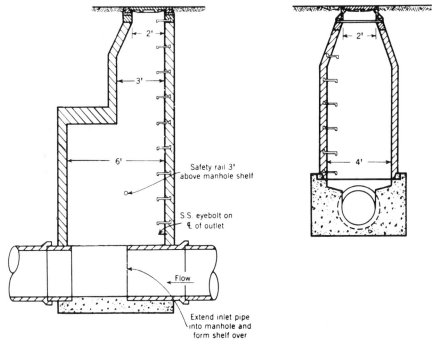

Fig. 7-3. Two manholes for intermediate size sewers (ft × 0.3 = m).

slope of the adjacent pipeline. If the channel is constructed properly this drop is unnecessary. The drop is, in fact, objectionable for it causes excessive turbulence just where it is least desirable and sacrifices head that might better be used toward the attainment of good slopes along the entire sewer. The usual practice calls for a continuation of the pipe slope through the manhole.

10. Manholes on Large Sewers

The operations and methods of maintenance in large sewers are not the same as in small ones, and manhole designs are specified accordingly. Sometimes a platform is provided at one side, or the manhole is simply a vertical shaft over the center of the sewer. In the latter case, a block of reinforced concrete is cast around the pipe and designed to form an adequate foundation to support the shaft plus transmitted traffic loads. Such manholes usually are built without steps since a man cannot step off into the water anyway. When entry is necessary, a workman is lowered in a chair hoist or cage. Large factory-made T-sections also can be used if adequate support is provided for the shaft and transmitted loads.

Where a sewer is larger than 600 mm (2 ft) and the small-sewer type of manhole is used, the diameter of the manhole should be increased sufficiently to maintain an adequate width of bench, preferably 0.3 m (1 ft) or more on each side. In sewers that a man cannot straddle, maintenance men frequently lay planks to bridge the channel. Hence, there must be adequate and well-formed benches on each side. Sometimes the entering pipe is extended 0.6 to 1 m (2 to 3 ft) into the manhole and mortared over to form a smooth platform, as shown for the larger manhole in Fig. 7-3.

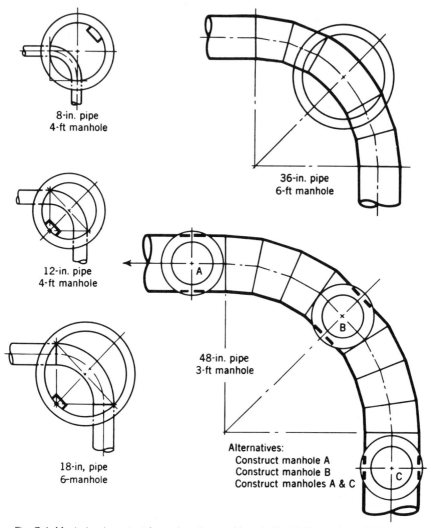

Fig. 7-4. Manhole placement for various types of bends (ft × 0.3 = m; in. × 2.54 = cm).

C. BENDS

Particular care must be used in the construction of curved channels to accommodate bends. The highest workmanship is necessary to produce channels that are smooth, with uniform sections, radius, and slope.

A curve of very short radius causes energy-wasting turbulence. Some authorities recommend that for optimum performance the radius of the centerline be three-times the pipe diameter or channel width. Reasonably satisfactory conditions usually can be obtained if the radius is not less than 1.5-times the diameter. If the velocity is supercritical, surface turbulence and energy losses arise even with long-radius bends.

The radius of curvature of a bend within a manhole is maximized if the points of tangency of the outer curve of the channel with the walls of the pipes are at the ends of the manhole diameter, as shown in Fig. 7-4 for the 300 and 450-mm (12 and 18-in.) pipes. This is true regardless of the angle. Bends of less than 90° can, of course, be accommodated more easily. For angles substantially less than 90° on sewers larger than 300 mm (12 in.) in diameter, the manhole is usually centered over the pipe.

Completion of a bend within a manhole is not necessary and becomes impossible as the pipe approaches the size of the manhole. Furthermore, when the size of the sewer is such that the manhole is only a chimney over the sewer, a manhole may be placed over the center of the curve, or on the downstream tangent, or perhaps two may be used, one upstream and one downstream. Fig. 7-4 shows some of the possible designs for manholes on bends.

Frequently extra fall in the channel invert is provided on bends to compensate for bend energy losses. When this is done, the extra fall should reflect only the expected losses. Although experimental data for large conduits are scarce, it would appear that for a well-made 90° bend with a centerline radius of curvature not less than one pipe diameter, the loss in an open channel should not exceed 0.4 of the velocity head. Thus, for a velocity of 1 m/sec (3 fps), the loss would probably be not more than 20 mm (0.06 ft) for subcritical flow. Energy loss will be greater with supercritical flow, but conservation of the energy of the stream is not likely to be important under that condition. In sewers where flows are small or velocities moderate, the energy losses at bends are usually ignored. The slight backing-up of the water due to the energy loss is usually inconsequential. (A more complete discussion of energy losses in bends, including those associated with supercritical flow, is found in Chapter 5.)

D. JUNCTIONS

On small sewers junctions are made in ordinary manholes, with the branch line curved into the main channel. Excessive widening of the main channel at the junction should be avoided. Eddying flows and accumulations of sludge and rags are the result of poor flow patterns prevailing in many junction manholes. To minimize these objectionable conditions, the invert of the branch lines may be brought in somewhat higher than the invert of the main channel where the two join. Channels are generally constructed in the manhole bottom for all lines. Sometimes right-angle junctions are used in small sewers. This usually causes less energy loss than the attempt to conserve some of the energy of the side stream by use of a curved channel.

Large junctions generally are constructed in cast-in-place reinforced concrete chambers entered through manhole shafts (their hydraulic design is discussed in Chapter 5).

E. DROP MANHOLES

If a sewer enters a manhole at an elevation considerably higher than the outgoing pipe, it is generally not satisfactory to let the stream merely pour into the manhole because the structure then does not provide an acceptable working space. Drop manholes are usually provided in these cases. Fig. 7-5 shows common types. These structures are not trouble-free. Sticks may bridge the

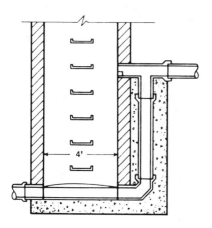 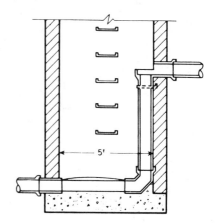

Fig. 7-5. Drop manholes (ft × 0.3 = m).

drop pipe, starting a stoppage. Because of such a stoppage or merely because of high flow, wastewater may spill out of the end of the pipe, making the manhole dangerous and objectionable as a place for a man to work. Cleaning equipment also may lodge in the drop pipe. Sometimes the drop pipe is placed inside the manhole, or the drop pipe is made of a larger diameter to minimize stoppages. Another arrangement sometimes used is to provide a cross instead of a tee outside the manhole, with the vertical pipe extended to the surface of the ground and with a suitable cover so that it is accessible for cleaning.

Drop manholes should be used sparingly and, generally, only when it is not economically feasible to steepen the incoming sewer. Some engineers eliminate drops by using vertical curves. It would appear that this should be a general rule for elevation differences of less than about 1 m (3 ft) and often for larger drops as well.

F. TERMINAL CLEANOUTS

Terminal cleanouts sometimes are used at the upstream ends of sewers, although most engineers now specify manholes. Their purpose is to provide a means for inserting cleaning tools, or for flushing, or for inserting an inspection light into the sewer.

A terminal cleanout consists of an upturned pipe coming to the surface of the ground. The turn should be made with bends so that flexible cleaning rods can be passed through it. The diameter should be the same as that of the sewer. The cleanout is capped with a cast-iron frame and cover (Fig. 7-6).

In the past, tees often were used instead of pipe bends and the structures were called "lampholes". Sewer cleaning equipment cannot be passed into the sewer through a tee, so such structures are not now considered to be good design practice.

Regulations in most areas allow terminal cleanouts (if at all) only within 45 to 60 m (150 to 200 ft) of a manhole.

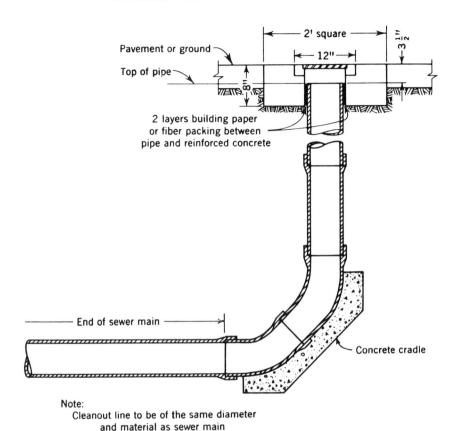

Fig. 7-6. Terminal cleanout (ft × 0.3 = m; in. × 2.54 = cm).

G. SERVICE LATERALS

Service laterals, also called house connections or service connections, are the branches between the street sewer and the property or curb lines, serving individual properties; they usually are required to be 100, 125 or 150 mm in diameter (4, 5 or 6 in.) preferably with a slope of 1 in 48, or 2%. Sometimes 1% slopes are allowed and this seems to serve just as well. If a stoppage occurs in a service lateral, it may be due to root penetration, grease, or sometimes corrosion (in the case of iron pipes). Steeper slopes are of no benefit in coping with those problems.

Materials, joints, and workmanship for service laterals should be equal to those of the street sewer to minimize infiltration and root penetration. Particular attention should be paid to the construction of service laterals, especially compaction of bedding and backfill material, and jointing techniques, since these sewers frequently represent the major source of infiltration/inflow in a sanitary sewer system (see Chapter 3).

Often building sewers are constructed to the property or curb line at the time the street sewer is constructed. To meet the future needs of unsubdivided properties, wyes or tees sometimes are installed at what are presumed to be convenient intervals. Laterals or stubs not placed in use should be plugged

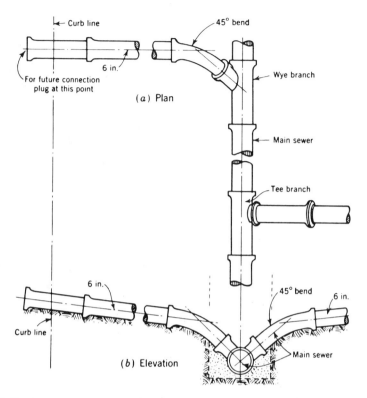

Fig. 7-7. Service connection for shallow sewer (in. × 2.54 = cm).

tightly. Fig. 7-7 shows typical connections.

If wyes or tees are not installed when the sewer is constructed, the sewer must be tapped later, and deplorably poor connections often have resulted. This is especially true for those connections made by breaking into the sewer and grouting-in a stub. Either a length of pipe should be removed and replaced with a wye or tee fitting, or better, a clean opening should be cut with proper equipment and a tee-saddle or tee-insert attached. Any connection other than to existing fittings must be made by experienced workmen under close supervision. Typical connections to a deep sewer are shown in Fig. 7-8.

In some places, test tees are required on the service lateral which permit the outlet to the street sewer to be plugged. This makes it possible to test the service lateral.

When a connection is to a concrete trunk sewer, a bell may be installed at the outside of the pipe. Three designs are shown in Fig. 7-9. Preferably the bell is provided by the manufacturer, but it can be installed in the field if necessary. It must be high enough on the pipe so that the lateral will not be flooded by high flows in the sewer.

Large trunks are not used ordinarily as collecting sewers. When they are over 900 to 1,200 mm (3 or 4 ft) in diameter, they frequently are paralleled by smaller sewers that enter the trunks at manholes.

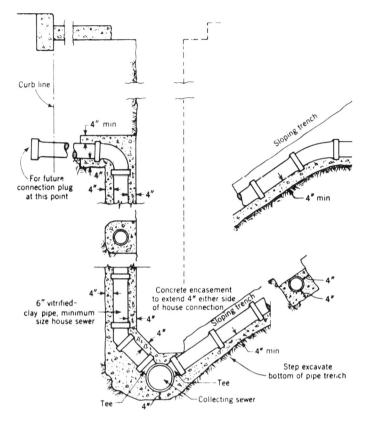

Fig. 7-8. Service connection for deep sewer (in. × 2.54 = cm).

H. CHECK VALVES AND RELIEF OVERFLOWS

Where the floor of a building is at an elevation lower than the top of the next upstream manhole on the sewer system, a stoppage in the main sewer can lead to overflow of wastewater into the building. Devices that sometimes are used to guard against such occurrences include backflow preventers or check valves and relief overflows.

Backflow preventers or check valves may be installed where the house plumbing discharges to the house sewers. Usually a double check valve is specified. Even so, such devices frequently do not remain effective over long periods of time.

Any overflow of wastewater is undesirable, but if a stoppage occurs in a street sewer, overflow may result. It will be from a manhole in the street, into a building, or on occasion at a designated overflow point. For the latter to be effective, it must be at an elevation lower than the floor level being protected. At this point, a relief device may be installed that encases a ball resting on a seat to close the end of a vertical riser and prevent flow into the sewer. The relief device must be constructed so that the ball will rise and allow overflowing wastewater to escape and thus provide relief. This practice is not encouraged except in the most extreme conditions and usually with regulatory approvals. Building owners who have valuable property in basements that might be flooded usually protect themselves, insofar as possible, by check valves.

142 GRAVITY SANITARY SEWER

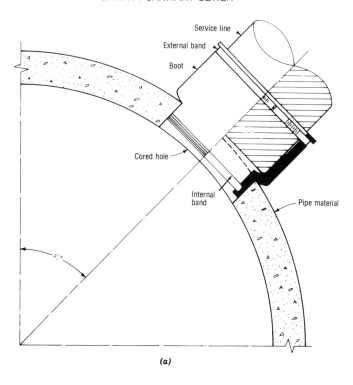

Fig. 7-9. a. Connections to large sewers.

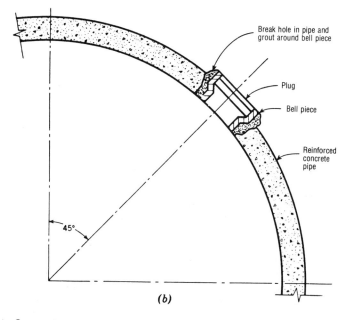

Fig. 7-9. b. Connections to large sewers.

APPURTENANCES AND SPECIAL STRUCTURES 143

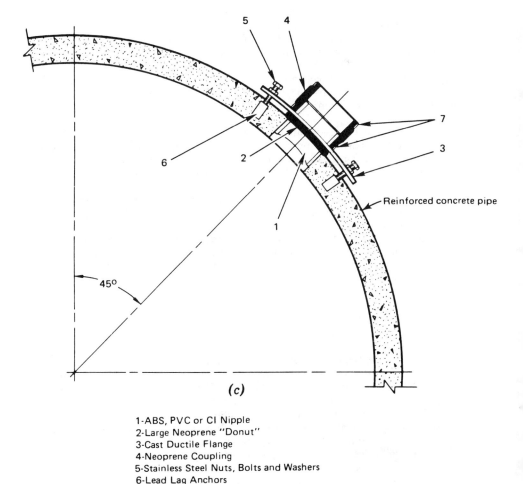

1-ABS, PVC or CI Nipple
2-Large Neoprene "Donut"
3-Cast Ductile Flange
4-Neoprene Coupling
5-Stainless Steel Nuts, Bolts and Washers
6-Lead Lag Anchors
7-Stainless Steel Clamps

Fig. 7-9. c. Connections to large sewers.

I. SIPHONS

"Siphon" in sewerage practice almost always refers to an inverted siphon or depressed sewer which would stand full even with no flow. Its purpose is to carry the flow under an obstruction such as a stream or depressed highway and to regain as much elevation as possible after the obstruction has been passed.

1. Single- and Multiple-Barrel Siphons

Siphons are often constructed with multiple barrels. The objective is to provide adequate self-cleansing velocities and maintenance flexibility under widely varying flow conditions. The primary barrel is designed so that a velocity of 0.6 to 1.0 m/sec (2 to 3 fps) will be reached at least once each day, even during the early years of operation. Additional pipes, regulated by lateral overflow weirs, assist progressively in carrying flows of greater magnitude, that is, maximum dry-weather flow to maximum storm flow. The overflow weirs may be considered as submerged obstacles, causing loss in head as flow passes over them. The weir losses may be assumed equal to the head necessary to produce critical velocity across the crest. Weir crest elevations are dependent on the depths of flow in the upstream sewer for the design quantity increments. Sample crest length calculations are presented in text books (1).

Many engineers maintain that for sanitary sewers, there is usually no need for multiple barrels. They reason that solids which settle out at low flows will flush out when higher flows are obtained, except for those heavy solids that would accumulate even at high flows. Single-barrel siphons generate less sulfide and cause less loss of hydraulic head than do multiple barrels. Single-barrel siphons have been built with diameters ranging from 150 to 2,300 mm (6 to 90 in.) or more. Engineers holding to this concept generally favor a small barrel if initial flows are to be much lower than in later years, with a larger barrel constructed at a later date or constructed at the outset and blocked off, so that it can always operate as a single-barrel siphon. In some situations a spare barrel may be desirable purely for emergency or maintenance use.

2. Profile

Two considerations which govern the profile of a siphon are provision for hydraulic losses and ease of cleaning.

The friction loss through the barrel will be determined by the design velocity. For calculating this head loss, it is advisable to use a conservative Hazen-Williams friction coefficient of 100 (Manning "n" from 0.014 for small sizes to 0.015 for the largest). In the case of multiple-barrel siphons, additional losses due to side-overflow weirs must be considered.

Siphons may need cleaning more often than gravity sewers. For easy cleaning by modern methods, the siphon should not have any sharp bends, either vertical or horizontal; only smooth curves of adequate radius should be used. The rising leg should not be so steep as to make it difficult to remove heavy solids by cleaning tools that operate hydraulically. Some agencies limit the rising slope to 15%, but slopes as great as 50% (30°) are used in some places. There should be no change of pipe diameter within the length of a barrel since this would hamper cleaning operations.

3. Air Jumpers

Positive pressure develops in the sewer atmosphere upstream from a siphon because of the downstream movement of air induced by the sewage flow. In extreme cases, this pressure may equal several inches (centimeters) of water. Air, therefore, tends to exhaust from the manhole at the siphon inlet, escaping in large amounts even from a pick hole. Under all except maximum flow conditions, there is a drop in water surface elevation into a siphon, with consequent turbulence and release of odors. The exiting air hence can be the cause of serious odor problems. Conversely, air is drawn in at the siphon outlet.

Attempts to close the inlet structure tightly will usually force the air out of plumbing vents or manholes farther upstream. Insofar as the attempt to close the sewer tightly is successful, oxygen depletion in the sewer atmosphere occurs, aggravating sulfide generation where this is a problem.

To overcome this difficulty, a number of siphons built in recent years have used air jumpers; that is, pipes that take the air off the top of the inlet structure and return it at the end of the siphon. Usually, the jumper pipe is a third to a half the diameter of the siphon. Sometimes the pipe can be suspended above the hydraulic grade line of the sewer, but in other cases it must run more or less parallel to the siphon. In these cases, provision must be made for dewatering the jumper, otherwise it will fill with condensate. In some cases a drain has been installed to a percolation pit. One large air jumper in use consists of 1,200 mm (48-in.) diameter pipe paralleling a 2,300 mm (90-in.) siphon, 610 m (2,000 ft) long, utilizing a sump pump for dewatering.

4. Sulfide Generation

Sulfide may be produced in a long siphon. There is nearly always a hydraulic jump or turbulence at a siphon inlet, which causes absorption of oxygen and delays the onset of sulfide buildup in comparison with pressure mains that lack this initial aeration. Thus, sulfide buildup may be delayed for as much as an hour if the wastewater is of low temperature or low BOD. When higher temperatures prevail, and especially if oxygen absorption at the inlet is minimal, sulfide buildup may be under way in 20 min. For any given flow and wastewater characteristics, the sulfide concentration produced in a filled pipe is roughly proportional to the pipe diameter. (For further details, see Chapter 4.)

J. FLAP OR BACKWATER GATES

Flap gates are installed at or near sewer outlets to prevent back-flooding of the sewer system by high tides or high stages in the receiving stream.

Flap gates may be made of wood, cast iron, or steel. They are commercially available in sizes up to 2.4 m (8 ft) in diameter. Larger gates can be fabricated from plates and structural shapes. They should be hinged by a link-type arrangment which allows the gate shutter to seat more securely. Hinge pins, linkages, and seats should be corrosion-resistant.

The maintenance of flap gates requires regular inspection and removal of debris from the pipe and outlet chamber, lubrication of hinge pins, and cleaning of seating surfaces.

K. SEWERS ABOVE GROUND

Occasionally, in rolling or hilly terrain, it is desirable and economical to build sewers above the surface of the ground or across gullies and stream valleys. Such sewers often are constructed in carefully compacted fill. Sometimes it is better to suspend a sewer over a waterway or a highway than to go under it by means of an inverted siphon. Sewer crossings in such cases have been constructed by installing or hanging the pipes on bridges, by fastening them to structural supports which rest on piers, by supporting them with suspension spans and cables, and by means of sewer pipe beams.

Structural design of suspended sewers is similar to that of comparable structures with supporting members of timber, steel, or reinforced concrete. Foundation piers or abutments should be designed to prevent overturning and settlement. The impact of flood waters and debris should be considered.

If the sewer is exposed, as on a trestle, steel pipe may be used, perhaps with coating and lining for corrosion protection. Sometimes sewers of other materials are carried inside steel pipes. The steel pipe may be supported by simple piers at suitable intervals.

In recent years, prestressed concrete pipe beams have been used to span waterways and other obstacles. Generally, they have been of three types:

(a) A rectangular section with a circular void extending the full length of member, either pretensioned or post-tensioned, and similar to a hollow box highway girder. This section normally is used for smaller sewers.

(b) A pretensioned circular pipe section which may be produced in most any diameter. This type is economical for long crossings.

(c) Reinforced concrete pipe sections assembled and post-tensioned to form the required sewer pipe beam. These beams may be fabricated economically using standard pipe forms and prestressing equipment. The pipes are cast with longitudinal cable ducts in the walls. After curing, the pipes are aligned, post-tensioning cables inserted, jacked to the design tension and anchored. Pressure grouting of the ducts completes the manufacture, and the sewer beam then is shipped to the job site for installation. Chapter 11 contains a picture of a prestressed concrete pipe beam constructed in this manner.

Protection against freezing and prevention of leakage are important design and construction considerations for above-ground sewers.

It has been found necessary in some designs to employ expansion jointing between above-ground and below-ground sewers. Special couplings are available for such purposes. Anchorage provision also must be made to prevent permanent creep. Expansion joints in sewers supported on bridges or buildings should match the expansion joints in the structures to which the sewer is attached.

L. UNDERWATER SEWERS AND OUTFALLS

1. Ocean Outfalls

Communities adjacent to the seacoast may discharge their treated wastewater into the ocean. Disinfected secondary effluents generally are discharged relatively close to shore but usually beyond the distance designated for body contact. Primary effluents are carried far enough to sea to avoid any undesirable effects. In the U.S., an attempt is being made to require that all ocean discharges have secondary treatment, but this is not the usual policy in other countries where the discharge of primary effluent is considered satisfactory if suitable depths and distances are reached.

For proper design, it is essential to obtain detailed data on the following:
(a) Profiles of possible outfall routes;
(b) nature of the ocean bottom;
(c) water density stratification or thermoclines, by seasons; and
(d) patterns of water movement at point of discharge and travel time to shore.

Since seawater is 2.5% denser than sanitary wastewater, the discharged wastewater rises rapidly, normally producing a "boil" at the surface. The rising plume mixes with a quantity of seawater which is generally from 10 to 100 or more times the wastewater flow. Dilution increases rapidly as the "wastewater field" moves away from the boil. The required length and depth of the outfall is related to the degree of treatment of the wastewater. The length must be calculated so that time and dilution will protect adequately the beneficial uses of the adjacent waters.

Much research has been done in recent years regarding the dilution of the wastewater and the die-away of the bacteria (2,3,4,5,6,7). A full treatment of that subject is beyond the scope of this manual.

Where the outfall is deep and there is good density stratification (thermocline), the rising plume may pick up enough cold bottom water so that the mixture is heavier than the surface water. The rising plume, therefore, stops beneath the surface, or reaches the surface and then resubmerges.

Diffusers may be used to gain maximum benefit of density stratification. If, however, they merely divide the flow into many small streams in a small area (a gas-burner type of diffuser) they do little good. The flow must be dispersed widely so that huge flows of dilution water can be utilized at low velocity.

The diffuser must be approximately level if it is to accomplish reasonably uniform distribution. For design of the diffuser, the rule of thumb may be used that the total cross-sectional area of the ports should not be more than half the cross-sectional area of the pipe. In large diffusers, often exceeding 1 km (0.6 miles) in length, the diffuser diameter may be stepped down in size toward the end (8). Computerized calculations are used in the design of these large diffusers.

These principles are well illustrated by the Los Angeles City outfall in Santa Monica Bay, California. The effluent is carried by a 3.7-m (12-ft) diam pipe to a point 8 km (5 miles) from shore at a depth of 58 m (190 ft), then dispersed through a Y-shaped diffuser with the two arms totalling 2,400 m (8,000 ft) in length. Except for certain periods in winter when the thermocline is practically nonexistent, no wastewater can be seen rising to the surface. The flow of effluent, essentially primary and unchlorinated, exceeds 13 m^3/sec (300 mgd), yet the bathing waters of the highly popular beaches on the Bay show no bacterial evidence of the wastewater discharge. The effluent outfall is supplemented by a second outfall 550 mm (22 in.) in diameter and 11 km (7 miles) long, which is used for disposal of digested sludge at a depth of 98 m (320 ft). The topography and currents at the point chosen for this discharge are such that there is no progressive accumulation of a sludge bank.

Outfalls into the open ocean generally are buried to a point where the water is deep enough to protect them from wave action, usually about 10 m (30 ft). Trenches in rock are formed by blasting. Beyond the buried portion the outfall rests on the bottom, with a flanking of rock to prevent currents from undercutting it where the bottom is soft.

Ocean outfalls in the smaller sizes are now usually made of steel pipe, mortar lined and coated. Steel pipes are welded and usually can be dragged into place from the shore, as was done with the 11-km (7-mile) sludge line of the City of Los Angeles. Relatively short pipelines are sometimes floated into place. Reinforced concrete pipe is used for the larger sizes. The joints for concrete outfalls usually are made with rubber gaskets similar to those used for construction on land. Special bolted restraints are used to secure the joints in small outfalls. Where the depth exceeds 10 m (30 ft), large pipes are simply laid on the bottom, sometimes with a rock cradle. They must be adequately flanked with rock. In this condition they are stable. Many of these outfalls are operating successfully.

2. Other Outlets

If effluent is discharged into an estuary or land-locked bay, special studies are needed to explore tidal currents, upstream flow of salt water at the bottom, available dilution, etc., in order to determine which discharge locations are compatible with various degrees of treatment.

Sewers discharging into streams with high-velocity flood flows require special thought in design to prevent undermining of the outlet structure as well as the pipe itself. Large outlets into the Mississippi River have been built with massive headwalls and wingwalls with deep foundations.

M. MEASURING WASTEWATER FLOWS

The design of a sewer system often requires measurements of flows in existing sewers, as well as the design of permanent monitoring facilities to measure and record future flows at one or more points in the system. There are many reasons why such measurements are made, two of which are to provide information needed in the administration of contracts between cooperating parties and to aid engineers in the planning of future expansions.

Two classes of metering devices may be distinguished: Those that are adapted to filled pipes, or "closed-channel flow", and those that make measurements in streams that have a surface exposed to the air as in a partly filled gravity sewer, or the "open-channel flow" condition.

A pressure main or force main from a pump station is a closed channel in which a Venturi-type meter, a sonic meter depending upon the Doppler effect, or a magnetic meter may be used. In rare cases the stream in a gravity sewer is caused to flow through a depressed reach as a pressure conduit where one of these devices has been placed. This section of the chapter, however, deals only with measurements of the open-channel type, since most sewer measurements are made by these methods.

There are several spot-check methods for determining flows. One is to measure the depth of flow and the velocity of the stream determined from using a dye or other indicator. Such a measurement is valid only if a reliable determination of average flow depth and velocity can be made. In large sewers, and especially if depths can be measured at several manholes, quite precise results can be obtained. At a flow of less than about 30 L/sec (1 cfs) in a sewer, the results are likely to be uncertain. Another spot-check method is the use of a tracer, usually NaCl, or a radioactive element, with subsequent sampling at a downstream point and analyses to determine the dilution of the tracer. It is possible to secure very precise results by this method, but only if the amount of tracer added is accurately measured and the downstream concentration is determined with a high degree of precision. It is a good method for calibrating any type of meter.

Often flow determinations can be made at a pump station by timing the filling of a wet well. If an average filling time is determined as well as an average time of pumping out, calculations can be made of the pumping rate. A time meter on the pump will then give a measure of total flow, but the pumping rate must be checked from time to time, since it may not remain constant.

The most common type of device for obtaining a continuous flow record in a gravity sewer is a weir, using that word in its general meaning of an obstruction over which the water flows. In engineering practice in the U.S., the word has come to be associated mostly with the sharp-crested weir used for measuring flow quantities, but a sharp crest is not universally implied. Streamlined weirs often have advantages over the sharp crest, especially for measuring raw wastewaters where the sharp crest may collect stringy matter.

In flowing over a weir, the stream passes through a "control section." For any given flow, the elevation of the water surface at the control section is

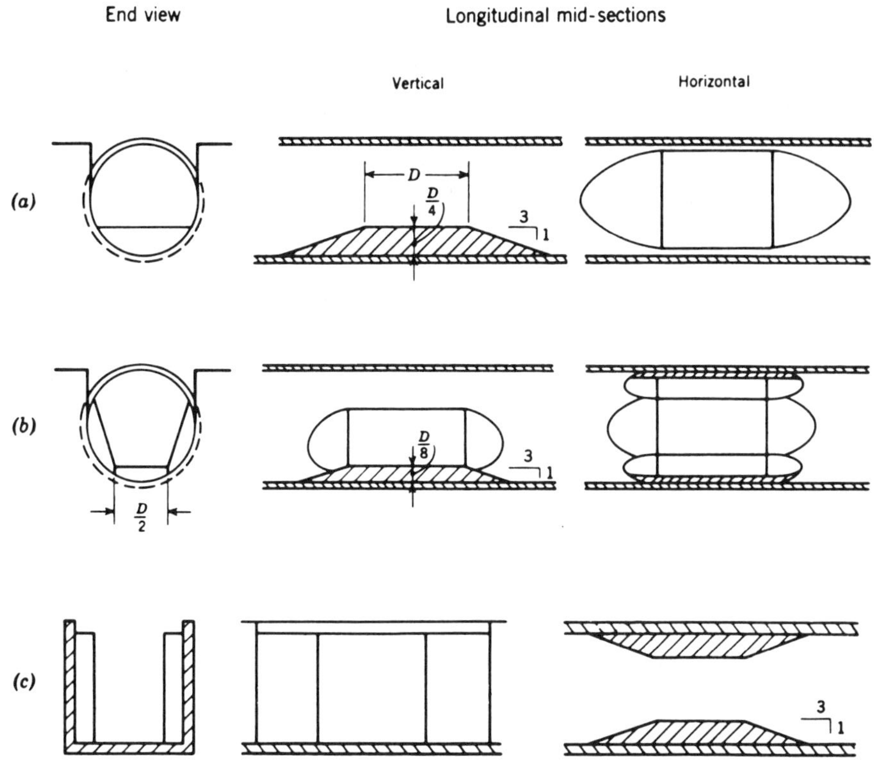

Fig. 7-10. Various shapes of Palmer-Bowlus flumes.

such that the total energy (elevation plus kinetic energy) is minimal. This section controls the upstream water elevation. The upstream elevation near the weir is used as a measure of the discharge over the weir, but the elevation reading needs to be increased by the amount of the kinetic energy of the velocity of the water at the place of measurement so as to obtain a figure for the total energy. Inherent in the calculations is the assumption that there is no energy loss between the point of measurement and the control section. Under extreme conditions this assumption may entail a significant error.

In 1936 H.K. Palmer and Fred Bowlus, both engineers at the Los Angeles County Sanitation Districts, showed the advantage of a streamlined weir (9). They devised a trapezoidal form which came to be known as a "Palmer-Bowlus flume" (Fig. 7-10b). (They described it as a Venturi flume, but it has little in common with the Venturi meters used for closed-channel measurements.) Palmer-Bowlus flumes are available commercially in various sizes. Later, Bowlus devised a simple slab form (unpublished work). Fig. 7-10a shows a Bowlus weir constructed for insertion into a 450-mm (18-in.) pipe. It is easily placed and easily retrieved by means of a chain attached to the upstream toe. Portable weirs of this type are used in sewers up to 675 mm (27 in.) in diameter, or larger if there is access by an opening larger than a standard manhole.

Since this type of weir can be placed in a sewer without altering the invert, it is particularly well suited where it is necessary to construct a metering station on an existing sewer. A Parshall flume, which is a widely used form of a streamlined weir with dimensions generally in arbitrarily fixed ratios, specifies a drop in the invert.

Ludwig and Ludwig (10) and Wells and Gotaas (11) experimented with both the trapezoidal and slab-type streamlined weirs. They found that these devices installed in sewers can meter flows up to 90% of the sewer capacity, and that the differences between actual flows and the flows shown by the theoretically calculated rating curves were less than 3%.

Streamlined weirs have often been built in channels of rectagular section, and in some of these the control section has been produced by merely constricting the sides, leaving the invert as a clear continuation of the invert of the rectangular channel (Fig. 7-10c). Calculation of the rating curve follows the same form as that for other rectangular streamlined weirs.

The theoretical equation for a rectangular streamlined weir is:

$$Q = \frac{2}{3} B \sqrt{\frac{2}{3} g H^3} \text{ (any fully consistent units)} \tag{7.1}$$

in which Q is the flow, B is the width at the control section, g is the gravitation constant, and H is the total energy of the stream, that is, the elevation relative to the elevation of the crest of the weir plus the kinetic energy at the same location. For a rectangular streamlined weir, the formula reduces to:

$$Q = 3.09 \, BH^{3/2} \text{ (cubic feet per second)} \tag{7.2a}$$

or $Q = 1.705 \, BH^{3/2}$ (cubic meters per second) (7.2b)

The discharge over a sharp-crested weir is a few percent greater than shown by the calculations in the preceeding equations because the control section over the sharp crest is slanted slightly upstream. The Parshall flume has a discharge about 7% greater than for a simple rectangular shape. Empirically determined rating curves have been published for Parshall flumes of various sizes.

Because of the wide utility of the Bowlus weir, a rating curve for it is shown in Table 7-1 for a 305-mm (1-ft) diameter pipe. The height of the weir is one-fourth of the pipe diameter. It is assumed that the weir is placed so that the critical section is in the outlet pipe from a manhole, and the depth measurement is made in a U-shaped channel in the manhole.

If the flow is very small it may be desirable to use a weir with a height of less than one-quarter the pipe diameter. The Palmer-Bowlus trapezoidal flume has a height equal to one-eighth diameter. If the velocity of approach is too great, a weir height greater than one-quarter diameter may be used. A computer can easily be programmed to provide a rating table for any chosen design.

The attainment of a critical flow condition over any type of weir is essential if a correct measurement is to be obtained. Usually, this condition is assured if a hydraulic jump is observed downstream. If the slope of the pipe is such that the normal flow of the water is supercritical, a hydraulic jump will be seen upstream from the weir instead of downstream, as a result of the retardation of the stream approaching the weir.

Table 7-1 Rating Curve for Streamlined Bowlus Weir Placed in Outlet Pipe of Manhole*

h or h/D (1)	Q, in cubic feet per second (2)	h, in feet (3)	Q, in cubic feet per second (4)	h, in feet (5)	Q, in cubic feet per second (6)
0.05	0.031				
0.06	0.041	0.26	0.43	0.46	1.09
0.07	0.052	0.27	0.45	0.47	1.13
0.08	0.064	0.28	0.48	0.48	1.17
0.09	0.077	0.29	0.51	0.49	1.21
0.10	0.091	0.30	0.54	0.50	1.25
0.11	0.106	0.31	0.57	0.51	1.29
0.12	0.121	0.32	0.60	0.52	1.33
0.13	0.138	0.33	0.63	0.53	1.37
0.14	0.155	0.34	0.66	0.54	1.41
0.15	0.174	0.35	0.70	0.55	1.45
0.16	0.193	0.36	0.73	0.56	1.49
0.17	0.213	0.37	0.76	0.57	1.53
0.18	0.233	0.38	0.80	0.58	1.58
0.19	0.255	0.39	0.83	0.59	1.62
0.20	0.277	0.40	0.87	0.60	1.66
0.21	0.300	0.41	0.90	0.61	1.70
0.22	0.324	0.42	0.94	0.62	1.74
0.23	0.348	0.43	0.98	0.63	1.79
0.24	0.373	0.44	1.01	0.64	1.83
0.25	0.399	0.45	1.05	0.65	1.87
				0.66	1.91

*Height of the weir is $D/4$, in which D is the pipe diameter in feet. The calculated flows, Q, are for a pipe 1 ft (305 mm) in diameter. For any other pipe size, use h/D in place of h, and multiply the flows by $D^{5/2}$

Note: 1 ft = 0.305 m; 1 cfs = 28.3 L/sec

Any kind of weir will cause some retardation of the upstream velocity. Therefore, there may be some stranding of solids upstream. Usually this is of no consequence, but if much grit accumulates it may materially increase the upstream velocity and reduce the apparent discharge. This effect is probably greater with a sharp-crested weir, and may be minimal with the Parshall flume and other weirs that do not have a raised bottom. A wooden shovel or paddle may be used to move excessive accumulations of sand over or through the streamlined weir.

Permanent metering stations are designed with vaults that provide access to the hydraulic device for maintenance and calibration.

The measurement of water elevation upstream from the weir or flume may be made with a streamlined float, provided that an arm or cords are used to hold it in place. Quite commonly a float well is provided outside the channel, sometimes outside the vault that houses the channel. The float well is connected to the channel with a pipe generally in the size range of 15 to 25 mm (0.5 to 1 in.) in diameter, terminating in a smooth opening flush with the wall of the channel. A major problem with float wells is the accumulation of solids. Fresh water is provided for periodic or continuous flushing of the well and connecting pipe. A drain is also useful in the bottom of the float well. Safety of the fresh water supply is protected by an air gap located higher than the highest water level under flooding conditions, which generally means higher than the ground surface. A leveling drain also should be provided, so arranged that when the connection to the channel is shut off and the leveling valve opened, the water will discharge to a sump or other low point until the level in the well exactly equals the elevation of the crest of the weir.

An alternative to the float well is a bubble tube. The pressure required to discharge air from the end of a pipe dipping into the water serves as a measure of the depth of water at that point. The end of the pipe is usually cut cleanly at a right angle to the axis of the pipe. The pipe may be perpendicular to the flow or it may be angled downstream so as to reduce the accumulation of paper and stringy material. Erroneous results will be obtained if the pipe angles upstream. Only a small air stream is needed. A bubble every 5 to 10 sec will suffice. Usually the air is supplied from a pressure tank recharged from time to time by a compressor. For temporary installations, a cylinder of air or carbon dioxide is a convenient source. Pressure sensors send signals that are converted to flow rates for indication and recording. Usually the recorder is outside the vault, to escape the humid, corrosive atmosphere in the vault.

N. REFERENCES

1. Fair, G. M., and Geyer, J. C., *Water Supply and Wastewater Disposal*, 1st Ed., John Wiley & Sons, Inc., New York City (1954)
2. Rawn, A. M. and Palmer, H. K., "Predetermining the Extent of a Sewage Field in Sea Water." *Trans. Amer. Soc. Civil Engr.*, 94, 1036 (1930)
3. Brooks, N. H., "Diffusion of Sewage Effluent in an Ocean Current." *Proc. 1st Conf. on Waste Disposal in Marine Environment*, Univ. Calif., Berkeley, Pergamon Press Ltd., London, England (1960)
4. Abraham, G., "Jet Diffusion in Stagnant Ambient Fluid," *Delft Inst. Hydr. Lab. Publ. No. 29*, Delft, The Netherlands (1963)
5. Pomeroy, R. D., "The Empirical Approach for Determining the Required Length of an Ocean Outfall," *Proc. 1st Conf. on Waste Disposal in Marine Environment*, Univ. of Calif., Berkeley, Pergamon Press Ltd., London, England (1960)
6. Pearson, E. A., "An Investigation of the Efficacy of Submarine Outfall Disposal of Sewage and Sludge," *Calif. Water Poll. Contr. Bd. Publ. No. 14*, Sacramento, Calif. (1956)
7. Gunnerson, C. G., "Marine Disposal of Wastes," *Jour. San. Eng. Div.*, Proc. Amer. Soc. Civil Engr. 87, SA1, 23 (1961)
8. Rawn, A. M., Bowerman, F. R., and Brooks, N. H., "Diffusers for Disposal of Sewage in Sea Water," *Jour. San. Eng. Div.*, Proc. Amer. Soc. Civil Engr., 86, SA2, 65 (1960)
9. Palmer, H. K., and Bowlus, F. D., "Adaption of Venturi Flumes to Flow Measurements in Conduits," *Trans. Amer. Soc. Civil. Engr.*, 101, 1195 (1936)
10. Ludwig, J. H., and Ludwig, R. G., "Design of Palmer-Bowlus Flumes," *Sewage and Industrial Wastes*, 23, 1096, (1951)
11. Wells, E. A., Jr., and Gotaas, H., "Design of Venturi Flumes in Circular Conduits," *Trans. Amer. Soc. Civil Engr.*, 123,749 (1958)
12. ASTM Standard C 923-79, "Standard Specification for Resilient Connectors Between Reinforced Concrete Manhole Structures and Pipes"

CHAPTER 8

MATERIALS FOR SEWER CONSTRUCTION

A. INTRODUCTION

Pipe for sanitary sewer construction generally is manufactured from various basic materials in accordance with nationally recognized product specifications. Each type of sanitary sewer pipe, its advantages and limitations, should be evaluated carefully in the selection of pipe materials for given applications.

Various factors are involved in the evaluation and selection of materials for sewer construction and are dependent on the anticipated conditions of service. Factors which may be involved and should be considered are:

(1) Intended use—type of wastewater
(2) Scour or abrasion conditions
(3) Installation requirements—pipe characteristics and sensitivities
(4) Corrosion conditions—chemical, biological
(5) Flow requirements—pipe size, velocity, slope and friction coefficient
(6) Infiltration/exfiltration requirements
(7) Product characteristics—pipe size, fitting and connection requirements, laying length
(8) Cost effectiveness—materials, installation, maintenance, life expectancy
(9) Physical properties—crush strength for rigid pipe, pipe stiffness or stiffness factor for flexible pipe, soil conditions, pipe beam loading strength, hoop strength for force main pipe, pipe shear loading strength, pipe flexural strength
(10) Handling requirements—weight, impact resistance

No single pipe product will provide optimum capability in every characteristic for all sanitary sewer design conditions. Specific application requirements should be evaluated prior to selecting or specifying pipe materials.

With the advancement of technology, new pipe materials are periodically being offered for use in sanitary sewer construction. Discussion of pipe materials provided in this chapter has been limited to the commonly accepted pipe materials currently available today for sanitary sewer applications. These products are listed below alphabetically within the two commonly accepted classifications of rigid pipe and flexible pipe:

(1) Rigid Pipe
    (a) Asbestos-cement pipe (ACP)
    (b) Cast iron pipe (CIP)
    (c) Concrete pipe
    (d) Vitrified clay pipe (VCP)

(2) Flexible Pipe
    (a) Ductile iron pipe (DI)
    (b) Steel pipe
    (c) Thermoplastic pipe
        • Acrylonitrile-butadiene-styrene (ABS)
        • ABS composite
        • Polyethylene (PE)

- Polyvinyl chloride (PVC)
(d) Thermoset plastic pipe
  - Reinforced plastic mortar (RPM)
  - Reinforced thermosetting resin (RTR)

B. SEWER PIPE MATERIALS

1. Rigid Pipe

Sanitary sewer pipe materials in this classification derive a substantial part of their basic earth load carrying capacity from the structural strength inherent in the rigid pipe wall. Commonly specified rigid sanitary sewer pipe materials are discussed in the following subsections.

a. Asbestos Cement Pipe (ACP)

Asbestos-cement pipe is used for both gravity and pressure sanitary sewers. The product, produced from asbestos fiber and cement, is available in nominal diameters from 100 through 900 mm (4 through 36 in.) and in some areas up to 1.05 m (42 in.). A full range of fittings is manufactured that is compatible with the class of pipe being used. Jointing is accomplished by compressing elastomeric rings between pipe ends and sleeves or couplings.

ACP, manufactured for gravity sanitary sewer applications, is available in seven strength classifications. The class designation represents the minimum crushing strength of the pipe expressed in pounds per linear foot of pipe.

Potential advantages of asbestos-cement pipe include:
- long laying lengths (in some situations),
- wide range of strength classifications,
- wide range of fittings available.

Potential disadvantages of asbestos-cement pipe include:
- subject to corrosion where acids are present,
- subject to shear and beam breakage when improperly bedded,
- low beam strength.

Asbestos-cement pipe is specified by pipe diameter and class or strength. The product should be manufactured in accordance with one or more of the following specifications:
- "Standard Specification for Asbestos-Cement Non-Pressure Pipe," ASTM C 428
- "Standard Specification for Asbestos-Cement Non-Pressure Small Diameter Sewer Pipe," ASTM C 644

b. Cast Iron Pipe (CIP)

CIP (gray iron) is used for both gravity and pressure sanitary sewers. Standards specify the product in nominal diameters from 50 mm through 1.2 m (2 through 48 in.) with a variety of joints. Cast iron fittings and appurtenances are generally available. Product availability is limited due to manufacturing conversion to ductile iron production.

CIP is manufactured in a number of thicknesses, classes, and strengths (5). A cement mortar lining with an asphaltic seal coating may be specified on the interior of the pipe. An exterior asphaltic coating is also commonly specified. Other linings and coatings may be specified.

Potential advantages of cast iron pipe (gray iron) include:
- long laying lengths (in some situations),

- high pressure and load bearing capacity.

Potential disadvantages of cast iron pipe (gray iron) include:
- subject to corrosion where acids are present,
- subject to chemical attack in corrosive soils,
- subject to shear and beam breakage when improperly bedded,
- high weight.

CIP is specified by nominal diameter, class, lining, and type of joint. CIP is manufactured in accordance with one or more of the following standard specifications:
- "Cast Iron Pipe Centrifugally Cast in Metal Molds, for Water or Other Liquids," ANSI A 21.6 (AWWA C 106)
- "Gray-Iron and Ductile Iron Fittings, 3 through 48-inch, for Water and Other Liquids," ANSI/AWWA C 110
- "Polyethylene Encasement for Gray and Ductile Iron Piping for Water and Other Liquids," ANSI/AWWA C 105/A 21.5
- "Flanged Cast-Iron and Ductile-Iron Pipe with Threaded Flanges," ANSI A 21.15 (AWWA C 115)
- "Cement Mortar Lining for Cast-Iron and Ductile-Iron Pipe and Fittings for Water," ANSI A 21.4 (AWWA C 104)

Additional information relative to the selection and design of CIP may be obtained from the *Ductile Iron Pipe Research Association Handbook – Ductile Iron Pipe, Cast Iron Pipe*.

Note: Cast iron soil pipe, produced for drain, waste and vent applications, is not used in construction of gravity or pressure sewer systems; however, the product may be used for house connections to gravity sewer systems.

c. Concrete Pipe

Reinforced and nonreinforced concrete pipe are used for gravity sanitary sewers. Reinforced concrete pressure pipe and prestressed concrete pressure pipes are used for pressure as well as sanitary sewers. Nonreinforced concrete pipe is available in nominal diameters from 100 through 900 mm (4 through 36 in.). Reinforced concrete pipe is available in nominal diameters from 300 mm through 5 m (12 through 200 in.). Pressure pipe is available in diameters from 300 mm through 3.0 m (12 through 120 in.). Concrete fittings and appurtenances such as wyes, tees, and manhole sections are generally available. A number of jointing methods are available depending on the tightness required and the operating pressure. Various linings and coatings are available.

A number of mechanical processes are used in the manufacture of concrete pipe. These processes use various techniques including centrifugation, vibration, packing and tamping for consolidating the concrete in forms. Gravity and pressure concrete pipe may be manufactured to any reasonable strength requirement by varying the wall thickness, concrete strength, quantity and configuration of reinforcing steel or prestressing elements (4).

Potential advantages of concrete pipe include:
- wide range of structural and pressure strengths,
- wide range of nominal diameters,
- wide range of laying lengths [generally 1.2 to 7.4 m (4 to 24 ft)].

Potential disadvantages of concrete pipe include:
- high weight,
- subject to corrosion where acids are present,
- subject to shear and beam breakage when improperly bedded.

Concrete pipe is normally specified by nominal diameter, class or D-load

strength and type of joint. The product should be manufactured in accordance with one or more of the following standard specifications:
- "Concrete Sewer, Storm Drain, and Culvert Pipe," ANSI/ASTM C 14
- "Reinforced Concrete Culvert, Storm Drain, and Sewer Pipe," ANSI/ASTM C 76
- "Reinforced Concrete Arch Culvert, Storm Drain, and Sewer Pipe," ANSI/ASTM C 506
- "Reinforced Concrete D-Load Culvert, Storm Drain, and Sewer Pipe," ANSI/ASTM C 655
- "Reinforced Concrete Elliptical Culvert, Storm Drain, and Sewer Pipe," ANSI/ASTM C 507
- "Reinforced Concrete Low-Head Pressure Pipe," ANSI/ASTM C 361
- "Joints for Circular Concrete Sewer and Culvert Pipe, Using Rubber Gaskets," ANSI/ASTM C 443
- "External Sealing Bands for Non-Circular Concrete Sewer, Storm Drain, and Culvert Pipe," ANSI/ASTM C 877

Additional information relative to the selection and design of concrete pipe may be obtained from the *American Concrete Pipe Association Concrete Pipe Design Manual* and *Concrete Pipe Handbook*.

d. Vitrified Clay Pipe (VCP)

VCP is used for gravity sanitary sewers. The product is manufactured from clay and shales. Clay pipe is vitrified at a temperature at which the clay mineral particles become fused. The product is available in diameters from 75 through 900 mm (3 through 36 in.) and in some areas up to 1.05 m (42 in.). Clay fittings are available to meet most requirements, with special fittings manufactured on request. A number of jointing methods are available (2).

VCP is manufactured in standard and extra-strength classifications, although in some areas the manufacture of standard-strength pipe is not common in sizes 300 mm (12 in.) and smaller. The strength of vitrified clay pipe varies with the diameter and strength classification. The pipe is manufactured in lengths up to 3 m (10 ft).

Potential advantages of vitrified clay pipe include:
- high resistance to chemical corrosion,
- high resistance to abrasion,
- wide range of fittings available.

Potential disadvantages of vitrified clay pipe include:
- limited range of sizes available,
- high weight,
- subject to shear and beam breakage when improperly bedded.

VCP is specified by nominal pipe diameter, strength, and type of joint. The product should be manufactured in accordance with one or more of the following standard specifications:
- "Standard Specification for Vitrified Clay Pipe, Extra Strength, Standard Strength and Perforated," ANSI/ASTM C 700
- "Compression Joints for Vitrified Clay Pipe and Fittings," ASTM C 425
- "Pipe, Clay, Sewer," Federal Specification SS-P-361d, Standard Methods of Testing Vitrified Clay Pipe, ANSI/ASTM 301

Additional information relative to the selection and design of vitrified clay pipe may be obtained from the *National Clay Pipe Institute Clay Pipe Engineering Manual*.

2. Flexible Pipe

Sanitary sewer pipe materials in this classification derive load carrying capacity from the interaction of the flexible pipe and the embedment soils effected by the deflection of the pipe to the point of equilibrium under load. Commonly specified flexible sanitary sewer pipe materials are discussed below.

a. Ductile Iron Pipe (DIP)

DIP is used for both gravity and pressure sanitary sewers. DIP is manufactured by adding cerium or magnesium to cast (gray) iron just prior to the pipe casting process. The product is available in nominal diameters from 75 mm through 1.35m (3 through 54 in.) and in lengths to 6.1 m (20 ft). Cast iron (gray iron) or ductile iron fittings are used with ductile iron pipe. Various jointing methods for the product are available.

DIP is manufactured in various thicknesses, classes and strengths. Linings for the interior of the pipe (e.g., cement mortar lining with asphaltic coating, coal tar epoxies, epoxies, polyethylene) may be specified. An exterior asphaltic coating and polyethylene exterior wrapping are also commonly specified (5).

Potential advantages of DIP include:
- long laying lengths (in some situations),
- high pressure and load bearing capacity,
- high impact strength,
- high beam strength.

Potential disadvantages of DIP include:
- subject to corrosion where acids are present,
- subject to chemical attack in corrosive soils,
- high weight.

DIP is specified by nominal diameter, class, lining, and type of joint. DIP should be manufactured in accordance with one or more of the following standard specifications:

- "Polyethylene Encasement for Gray and Ductile Cast-Iron Piping for Water and Other Liquids," ANSI A 21.5 (AWWA C 105)
- "Ductile Iron Gravity Sewer Pipe," ASTM A 746
- "Gray-Iron and Ductile Iron Fittings, 3 inch through 48 inch, for Water and Other Liquids," ANSI/AWWA C 110
- "Cement Mortar Lining for Cast-Iron and Ductile-Iron Pipe and Fittings for Water," ANSI A 21.4 (AWWA C 104)

Additional information relative to the selection and design of DIP may be obtained from the *Ductile Iron Pipe Research Association Handbook – Ductile Iron Pipe, Cast Iron Pipe.*

b. Steel Pipe

Steel pipe is rarely used for sanitary sewers; when used, it usually is specified with interior protective coatings or linings (polymeric, bituminous, asbestos, etc.). Steel pipe is fabricated in diameters from 200 mm through 3.0 m (8 through 120 in.). Appurtenances include tees, wyes, elbows, and manholes fabricated from steel. Various linings and coatings are available (1).

Steel is generally manufactured in lengths up to 12.2 m (40 ft).

Potential advantages of steel pipe include:
- light weight,
- long laying lengths (in some situations).

Potential disadvantages of steel pipe include:

- subject to corrosion where acids are present,
- subject to chemical attack in corrosive soils,
- difficulty in making lateral connections,
- poor hydraulic coefficient (unlined corrugated steel pipe),
- subject to excessive deflection when improperly bedded or haunched,
- subject to turbulence abrasion.

Steel pipe is specified by size, shape, wall profile, gauge or wall thickness and protective coating or lining. Steel pipe should be manufactured in accordance with one or more of the following standard specifications:

- "Pipe, Corrugated (Iron or Steel, Zinc Coated)," Federal Specification WW-P-405
- "Corrugated Metal Culvert Pipe," AASHTO M-36
- "General Requirements for Delivery of Zinc Coated (Galvanized) Iron or Steel Sheets, Coils and Cut Lengths Coated by the Hot Dip Method," ASTM A 475
- "Zinc-Coated (Galvanized) Corrugated Iron or Steel Culverts and Underdrains," AASHTO M-36
- "Precoated, Galvanized Steel Culverts and Underdrains," AASHTO M-245
- "Precoated (Polymeric) Galvanized Steel Sewer and Drainage Pipe," ASTM A 762
- "Pipe, Corrugated Steel, Zinc Coated (Galvanized)," ASTM A 760

Additional information relative to the selection and design of steel pipe may be obtained from the *American Iron and Steel Institute Steel Highway Construction and Drainage Handbook* and the *Sewer Manual for Corrugated Steel Pipe* prepared by the National Corrugated Steel Pipe Association.

c. Thermoplastic Pipe

Thermoplastic materials include a broad variety of plastics that can be repeatedly softened by heating and hardened by cooling through a temperature range characteristic for each specific plastic. Thermoplastic pipe product design should be based on long-term data. Generally, thermoplastic materials used in sanitary sewers are limited to acrylonitrile-butadiene-styrene (ABS), polyethylene (PE), and polyvinyl chloride (PVC).

1. Acrylonitrile-Butadiene-Styrene (ABS) Pipe

ABS pipe is used for both gravity and pressure sanitary sewers. Nonpressure rated ABS sewer pipe is available in nominal diameters from 75 through 300 mm (3 through 12 in.) and in lengths up to 11.2 m (35 ft). A variety of ABS fittings and several jointing systems are available.

ABS pipe is manufactured by extrusion of ABS plastic material.

ABS gravity sanitary sewer pipe is available in three dimension ratio (DR) classifications (23.5, 35 and 42) depending on nominal diameter selected. The classifications relate to three pipe stiffness values, PS 1000, 300, and 150 kPa (PS 150, 45 and 20 psi), respectively. The DR is the ratio of the average outside diameter to the minimum wall thickness of the pipe.

Potential advantages of ABS pipe include:
- light weight,
- long laying lengths (in some situations),
- high impact strength,
- ease in field cutting and tapping.

Potential disadvantages of ABS pipe include:
- limited range of sizes available,
- subject to environmental stress cracking,

- subject to excessive deflection when improperly bedded and haunched,
- subject to attack by certain organic chemicals,
- subject to surface change effected by long-term ultra-violet exposure.

ABS pipe is specified by nominal diameter, dimension ratio, pipe stiffness and type of joint. ABS pipe should be manufactured in accordance with one or more of the following specifications:
- "Acrylonitrile-Butadiene-Styrene (ABS) Sewer Pipe and Fittings," ANSI/ASTM D 2751
- "Solvent Cement for Acrylonitrile-Butadiene-Styrene (ABS) Plastic Pipe and Fittings," ANSI/ASTM D 2235
- "Elastomeric Seals (Gaskets) for Joining Plastic Pipe," ANSI/ASTM F 477
- "Joints for Drain and Sewer Plastic Pipes Using Flexible Elastomeric Seals," ANSI/ASTM D 3212
- "PVC and ABS Injected Solvent Cemented Plastic Pipe Joints,"; ANSI/ASTM F 545

2. Acrylonitrile-Butadiene-Styrene (ABS) Composite Pipe

ABS composite pipe is used for gravity sanitary sewers. The product is available in nominal diameters from 200 through 375 mm (8 through 15 in.) and in lengths from 2 to 4 m (6.25 to 12.5 ft). ABS fittings are available for the product. The jointing systems available include elastomeric gasket joints and solvent cemented joints.

ABS composite pipe is manufactured by extrusion of ABS plastic material with a series of truss annuli which are filled with filler material such as light-weight portland cement concrete.

Potential advantages of ABS composite pipe include:
- light weight,
- long laying lengths (in some situations),
- ease in field cutting.

Potential disadvantages of ABS composite pipe include:
- limited range of sizes available,
- subject to environmental stress cracking,
- subject to rupture when improperly bedded,
- subject to attack by certain organic chemicals,
- subject to surface change effected by long-term ultra-violet exposure.

ABS composite pipe is specified by nominal diameter and type of joint. ABS composite pipe should be manufactured in accordance with one or more of the following standard specifications:
- "Acrylonitrile-Butadiene-Styrene (ABS) Composite Sewer Piping," ANSI/ASTM D 2680
- "Solvent Cement for Acrylonitrile-Butadiene-Styrene (ABS) Plastic Pipe and Fittings," ANSI/ASTM D 2235
- "Joints for Drain and Sewer Plastic Pipes Using Flexible Elastomeric Seals," ANSI/ASTM D 3212
- "Elastomeric Seals (Gaskets) for Joining Plastic Pipe," ANSI/ASTM F477

3. Polyethylene (PE) Pipe

PE pipe is used for both gravity and pressure sanitary sewers. Nonpressure PE pipe, primarily used for sewer relining, is available in nominal diameters from 100 mm through 1.2 m (4 through 48 in.). PE fittings are available. Jointing is primarily accomplished by butt-fusion or flanged adapters.

PE pipe is manufactured by extrusion of PE plastic material. Nonpressure PE pipe is produced at this time in accordance with individual manufacturer's product standards.

Potential advantages of PE pipe include:
- long laying lengths (in some situations),
- light weight,
- high impact strength,
- ease in field cutting.

Potential disadvantages of PE pipe include:
- relatively low tensile strength and pipe stiffness,
- limited range of sizes available,
- subject to environmental stress cracking,
- subject to excessive deflection when improperly bedded and haunched,
- subject to attack by certain organic chemicals,
- subject to surface change effected by long-term ultra-violet exposure,
- special tooling required for fusing joints.

PE pipe is specified by material designation nominal diameter (inside or outside), standard dimension ratios and type joint. PE pipe should be manufactured in accordance with one or more of the following specifications:
- "Butt Heat Fusion Polyethylene (PE) Plastic Fittings for Polyethylene (PE) Plastic Fittings for Polyethylene (PE) Pipe and Tubing," ANSI/ASTM D 3261
- "Polyethylene (PE) Plastic Pipe (SDR-PR)," ANSI/ASTM D 2239
- "Polyethylene (PE) Plastic Pipe (SDR-PR) Based on Controlled Outside Diameter," ASTM D 3035

4. Polyvinyl Chloride (PVC) Pipe

PVC pipe is used for both gravity and pressure sanitary sewers. Nonpressure PVC sewer pipe is available in nominal diameters from 100 through 675 mm (4 through 27 in.). PVC pressure and nonpressure fittings are available. PVC pipe is generally available in lengths up to 6.1 m (20 ft). Jointing is primarily accomplished with elastomeric seal gasket joints, although solvent cement joints for special applications are available (6).

PVC pipe is manufactured by extrusion of the plastic material.

Nonpressure PVC sanitary sewer pipe is provided in two dimension ratios (DR 35 and 41) which relate to two pipe stiffness values, PS 320 and 190 kPa (PS 46 and 28 psi), respectively.

Potential advantages of PVC pipe include:
- light weight,
- long laying lengths (in some situations),
- high impact strength,
- ease in field cutting and tapping.

Potential disadvantages of PVC pipe include:
- subject to attack by certain organic chemicals,
- subject to excessive deflection when improperly bedded and haunched,
- limited range of sizes available,
- subject to surface changes effected by long-term ultra-violet exposure.

PVC pipe is specified by nominal diameter, dimension ratio, pipe stiffness and type of joint. PVC pipe should be manufactured in accordance with one or more of the following standard specifications:
- "Type PSM Poly(Vinyl Chloride) (PVC) Sewer Pipe and Fittings,"

ANSI/ASTM D 3034
- "Type PSP Poly(Vinyl Chloride) (PVC) Sewer Pipe and Fittings," ANSI/ASTM D 3033
- "Elastomeric Seals (Gaskets) for Joining Plastic Pipe," ANSI/ASTM F 477
- "Joints for Drain and Sewer Plastic Pipes Using Flexible Elastomeric Seals," ANSI/ASTM D 3212
- "PVC and ABS Injected Solvent Cemented Plastic Pipe Joints," ANSI/ASTM F 545
- "Solvent Cements for Poly(Vinyl Chloride) (PVC) Plastic Pipe and Fittings," ANSI/ASTM D 2564
- "Standard Specification for Polyvinyl Chloride (PVC) Large Diameter Plastic Gravity Sewer Pipe and Fittings," ASTM F-679

Additional information relative to the selection and design of PVC pipe may be obtained form the *Uni-Bell Handbook of PVC Pipe – Design and Construction.*

d. Thermoset Plastic Pipe

Thermoset plastic materials include a broad variety of plastics. These plastics, after having been cured by heat or other means, are substantially infusible and insoluble. Thermoset plastic pipe product design should be based on long-term data. Generally, thermoset plastic materials used in sanitary sewers are provided in two categories—reinforced thermosetting resin (RTR) and reinforced plastic mortar (RPM).

1. Reinforced Thermosetting Resin (RTR) Pipe

RTR pipe is used for both gravity and pressure sanitary sewers. RTR pipe is generally available in nominal diameters from 25 through 300 mm (1 through 12 in.) manufactured in accordance with ASTM standard specifications. The product is available in nominal diameters from 300 mm through 3.6 m (12 through 144 in.) manufactured in accordance with individual manufacturer's specifications. In small diameters, RTR fittings are available. In larger diameters RTR fittings are manufactured as required. A number of jointing methods are available. Various methods of interior protection (e.g., thermoplastic or thermosetting liners or coatings) are available.

RTR pipe is manufactured using a number of methods including centrifugal casting, pressure laminating and filament winding. In general, the product contains fibrous reinforcement materials, such as fiberglass, embedded in or surrounded by cured thermosetting resin.

Potential advantages of RTR pipe include:
- light weight,
- long laying lengths (in some situations).

Potential disadvantages of RTR pipe include:
- subject to strain corrosion in some environments,
- subject to excessive deflection when improperly bedded and haunched,
- subject to attack by certain organic chemicals,
- subject to surface change effected by long-term ultra-violet exposure.

RTR pipe is specified by nominal diameter, pipe stiffness, lining and coating, method of manufacture, thermoset plastic material, and type of joints. RTR pipe should be manufactured in accordance with one or more of the following standard specifications:
- "Filament-Wound Reinforced Thermosetting Resin Pipe," ASTM D 2996

- "Centrifugally Cast Reinforced Thermosetting Resin Pipe," ANSI/ASTM D 2997
- "Machine-Made Reinforced Thermosetting Resin Pipe," ASTM D 2310

2. Reinforced Plastic Mortar (RPM) Pipe

RPM pipe is used for both gravity and pressure sewers. RPM pipe is available in nominal diameters from 200 mm through 3.6 m (8 through 144 in.). In smaller diameters, RPM fittings are generally available. In larger diameters, RPM fittings are manufactured as required. A number of jointing methods are available. Various methods of interior protection (e.g., thermoplastic or thermosetting liners or coatings) are available.

RPM pipe is manufactured containing fibrous reinforcements such as fiberglass and aggregates such as sand embedded in or surrounded by cured thermosetting resin.

Potential advantages of RPM pipe include:
- light weight,
- long laying lengths (in some situations).

Potential disadvantages of RPM pipe include:
- subject to strain corrosion in some environments,
- subject to excessive deflection when improperly bedded and hanched,
- subject to attack by certain organic chemicals,
- subject to surface change effected by long-term ultra-violet exposure.

RPM pipe is specified by nominal diameter, pipe stiffness, stiffness factor, beam strength, hoop tensile strength, lining or coating, thermoset plastic material, and type of joint. RPM pipe should be manufactured in accordance with one or more of the following standard specifications:
- "Reinforced Plastic Mortar Sewer Pipe," ANSI/ASTM D 3262
- "Reinforced Plastic Mortar Sewer and Industrial Pressure Pipe," ASTM D 3754

C. PIPE JOINTS

1. General Information

The requirement for the control of ground water infiltration and wastewater exfiltration in sanitary sewer systems renders the specification of pipe joint design essential to proper sanitary sewer design. A substantial variety of pipe joints is available for the different pipe materials used in sanitary sewer construction. A common requirement which must be imposed on the design of all sanitary sewer systems, regardless of the type of sewer pipe specified, is the use of reliable, tight pipe joints. A good pipe joint must be watertight, root resistant, flexible and durable. In general, today, various forms of gasket (elastomeric seal) pipe joints are used in sanitary sewer construction. They generally can be assembled by unskilled labor in a broad range of weather conditions and environments with good assurance of a reliable, tight seal preventing leakage.

In common practice, water infiltration/exfiltration testing or air exfiltration testing is specified for typical nonpressure sanitary sewer system construction to demonstrate that infiltration/exfiltration is within acceptable limits. (The subject of infiltration/exfiltration is discussed in Chapter 6.)

2. Types of Pipe Joints

Commonly specified sanitary sewer pipe joints include:

a. Gasket Pipe Joints

Gasket joints effect a seal against leakage through compression of an elastomeric seal or ring. Gasket pipe joint design is generally divided into two types—push-on pipe joint and mechanical compression pipe joint.

1. Push-on Pipe Joint

This type of pipe joint uses a continuous elastomeric ring gasket which is compressed into an annular space formed by the pipe, fitting or coupler socket and the spigot end of the pipe providing a positive seal when the pipe spigot is pushed into the socket. When using this type of pipe joint in pressure sanitary sewers, thrust restraint may be required to prevent joint separation under pressure. Push-on pipe joints (fittings, couplers, or integral bells) are available on nearly all pipe products mentioned.

2. Mechanical Compression Pipe Joint

This type of pipe joint uses a continuous elastomeric ring gasket which provides a positive seal when the gasket is compressed by means of mechanical device. When using this type of pipe joint in pressure sanitary sewers, thrust restraint may be required to prevent joint separation under pressure. This type of pipe joint may be provided as an integral part of cast iron or ductile iron pipe. When incorporated into a coupler, this type of pipe joint may be used to join two similarly sized plain spigot ends of any commonly used sewer pipe materials.

b. Bituminous Pipe Joints

This type of pipe joint involves use of hot-poured or cold packed bituminous material forced into a bell-and-spigot pipe joint to provide a seal. The use of this joint is discouraged in that reliable, watertight joints are not assured.

c. Cement Mortar Pipe Joint

This type of pipe joint involves use of shrink-compensating cement mortar placed into a bell-and-spigot pipe joint to provide a seal. The use of this joint is discouraged in that reliable, watertight joints are not assured. Cement mortar joints are not flexible and may crack due to any pipe movement.

d. Elastomeric Sealing Compound Pipe Joints

Elastomeric sealing compound may be used in jointing properly prepared concrete gravity sanitary sewer pipe. Pipe ends must be sandblasted and primed for elastomeric sealant application. The sealant, a thixotropic, two-compound elastomer, is mixed on the job site and applied with a caulking gun and spatula. The pipe joint, when assembled with proper materials and procedures, provides a positive seal against leakage in gravity sewer pipe.

e. Solvent Cement Pipe Joints

Solvent cement pipe joints may be used in jointing thermoplastic pipe materials such as ABS, ABS composite, and PVC pipe. This type of pipe joint involves bonding a sewer pipe spigot into a sewer pipe bell or coupler using a solvent cement. Solvent cement joints can provide a positive seal provided the proper cement is applied under proper ambient conditions with proper techniques. Reference should be made to ASTM D 402 for safe handling procedures. Required precautions should be taken to assure adequate trench ventilation and protection for workers installing the pipe. Solvent cement pipe joints may be desired in special situations and with some plastic fittings.

f. Heat Fusion Pipe Joints

Heat fusion pipe joints are commonly specified for PE sanitary sewer pipe. The general method of jointing PE sanitary sewer pipe involves butt fusion of the pipe lengths, end to end. After the ends of two lengths of PE pipe are trimmed and softened to a melted state with heated metal plates, the pipe ends are forced together to the point of butt fusion, providing a positive seal (9). The pipe joint does not require thrust restraint in pressure applications. Trained technicians with special apparatus are required to achieve reliable watertight pipe joints.

g. Mastic Pipe Joints

Mastic pipe joints are frequently used for special non-round shapes of concrete gravity sewer pipe which are not adaptable for use with gasket pipe joints. The mastic material is placed into the annular space to provide a positive seal. Application may be by trowelling, caulking, or by the use of preformed segments of mastic material in a manner similar to gaskets. Satisfactory performance of the pipe joints depends upon the proper selection of primer, mastic material, and good workmanship in application.

h. Sealing Band Joints

External sealing bands of rubber made in conformance with ASTM C 877 are also used on noncircular concrete sewer pipe. These elastomeric bands are wrapped tightly around the exterior of the pipe at the joint and extend several inches (centimeters) on each side of the joint. Sealing against the concrete is achieved by a mastic applied to one side of the band.

D. SUMMARY

With consideration of appropriate factors essential to the proper design of sanitary sewer systems, it is apparent that the broad variety of materials for sanitary sewer construction available is a source of significant benefit. Each sewer pipe product and each type of pipe joint offer distinct advantages which can prove to be specifically beneficial in given applications. Each sewer pipe product and each type of pipe joint also present limitations which must be understood and can generally be accommodated in proper system design for given applications.

E. REFERENCES

1. *AISI Steel Highway Construction and Drainage Products Handbook,* American Iron and Steel Institute, Washington, D.C.
2. *Clay Pipe Engineering Manual,* National Clay Pipe Institute, Washington, D.C. (1978)
3. *Concrete Pipe Design Manual,* American Concrete Pipe Association, Vienna, Va. (1980)
4. *Concrete Pipe Handbook,* American Concrete Pipe Association, Vienna, Va. (1980)
5. *Handbook – Ductile Iron Pipe, Cast Iron Pipe,* Ductile Iron Pipe Research Association, Oak Brook, Ill. (1978)
6. *Handbook of PVC Pipe – Design and Construction* Uni-Bell Plastic Pipe Association, Dallas, Tex. (1982)
7. "Recommended Standards for Sewage Works," Committee of the Great Lakes-Upper Mississippi River Board of State Sanitary Engineers. (1978)

8. "Safe Handling of Solvent Cements Used for Joining Thermoplastic Pipe and Fittings," ASTM F-402, American Society for Testing & Materials, Philadelphia, Pa. (1980)
9. "Underground Installation of Polyethylene Piping," Plastics Pipe Institute, New York, N.Y.
10. "Wear Data of Different Pipe Materials at Sewer Pipelines," Institute for Hydromechanic and Hydraulic Structures, Technical University of Darmstadt, Darmstadt, West Germany (May 1973)

CHAPTER 9
STRUCTURAL REQUIREMENTS

A. INTRODUCTION

The structural design of a sanitary sewer requires that the supporting strength of the installed sewer pipe, divided by a suitable factor of safety, must equal or exceed the loads imposed on it by the combined weight of soil and any superimposed loads.

This chapter presents generally accepted criteria and methods for determining combined loads and supporting strength of the sewer pipe, as well as procedures for combining these elements with the application of a factor of safety to produce a safe and economical design.

Methods are presented for estimating probable maximum loads caused by soil forces and for both static and moving superimposed loads. Where so noted, the methods apply to rigid and flexible conduits in the three most common conditions of installation: In a trench in natural ground; in an embankment; and in a tunnel.

The design of rigid and flexible pipes is treated separately. There are no specific design procedures given for flexible pipes of intermediate stiffness. For such cases, design procedures such as computer analysis based on soil-structure interaction or the designs for rigid or flexible pipes may be used (not interchangeably) for conservative results.

The supporting strength of a buried sewer pipe is a function of installation conditions as well as the strength of the sewer pipe itself. Structural analysis and design of the sewer line are problems of soil-structure interaction. This chapter presents procedures for determining the field or installed supporting strength of rigid sewer pipe based on its established relationship to the laboratory test strength. It also presents methods of predicting approximate field deflections for flexible pipe, based on empirical methods.

Since installation conditions have such an important effect on both load and supporting strength, a satisfactory sewer construction project requires accurate assumed design conditions from the job site. Therefore, this chapter also includes a section on recommendations for construction and field observations to achieve this goal.

This chapter does not include information on reinforced concrete design of rigid sewer pipe sections. Reference should be made to standard textbooks and to ACI/ASTM Specifications or industry handbooks for such design data.

B. LOADS ON SEWERS CAUSED BY GRAVITY EARTH FORCES

1. General Method — Marston Theory

Marston developed methods for determining the vertical load on buried conduits caused by soil forces in all of the most commonly encountered construction conditions (1,2). His methods are based on both theory and experiment and have achieved acceptance as being useful and reliable.

More recent analysis and actual observation of field performance have shown that designs based on the Marston Theory yield satisfactory results especially for small diameter conduits in narrow trenches. For larger diameter conduits the results are conservative.

In general, the theory states that the load on a buried pipe is equal to the weight of the prism of soil directly over it, called the interior prism, plus or minus the frictional shearing forces transferred to that prism by the adjacent prisms of soil — the magnitude and direction of these frictional forces being a function of the relative settlement between the interior and adjacent soil prisms. The theory makes the following assumptions:

(a) The calculated load is the load which will develop when ultimate settlement has taken place.

(b) The magnitude of the lateral pressures which induce the shearing forces between the interior and adjacent soil prisms is computed in accordance with Rankine's theory.

(c) Cohesion is negligible except for tunnel conditions.

The general form of Marston's equation is:

$$W = CwB^2 \tag{9.1}$$

in which W is the vertical load per unit length acting on the sewer pipe because of gravity soil loads; w is the unit weight of soil; B is the trench width or sewer pipe width, depending on installation conditions; and C is a dimensionless coefficient that measures the effect of the following variables:
- The ratio of the height of fill to width of trench or sewer pipe;
- the shearing forces between interior and adjacent soil prisms;
- the direction and amount of relative settlement between interior and adjacent soil prisms for embankment conditions.

2. Types of Loading Conditions

Although the general form of Marston's equation includes all the factors necessary to analyze all types of installation conditions, it is convenient to classify these conditions, write a specialized form of the equation, and prepare separate graphs and tables of coefficients for each.

The accepted system of classification is shown diagrammatically in Fig. 9-1 and is described briefly below.

Trench conditions are defined as those in which the sewer pipe is installed in a relatively narrow trench cut in undisturbed ground and covered with soil backfill to the original ground surface.

Embankment conditions are defined as those in which the sewer pipe is covered above the original ground surface or when a trench in undisturbed soil is so wide that trench wall friction does not affect the load on the sewer pipe. The embankment classification is further subdivided into two major subclassifications — positive projecting and negative projecting. Sewer pipe is defined as positive projecting when the top of the sewer pipe is above the adjacent original ground surface. Negative projecting sewer pipe is that installed with the top of the sewer pipe below the adjacent original ground surface in a trench which is narrow with respect to the size of pipe and the depth of cover (Fig. 9-1), and when the native material is of sufficient strength that the trench shape can be maintained dependably during placement of the embankment.

A special case, called the induced trench condition, may be employed to minimize the load on a conduit under an embankment of unusual height.

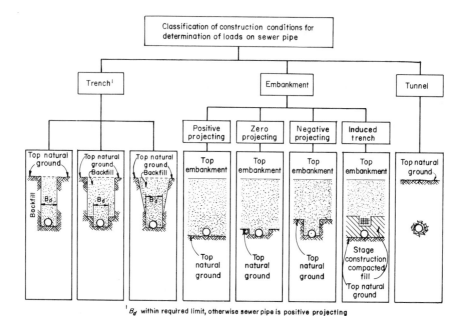

Fig. 9-1. Classification of construction conditions.

3. Loads for Trench Conditions

Sewers usually are constructed in ditches or trenches which are excavated in natural or undisturbed soil, and then covered by refilling the trench to the original ground line. This construction procedure often is referred to as "cut and cover," "cut and fill," or "open cut."

The vertical soil load to which a sewer pipe in a trench is subjected is the resultant of two major forces. The first is produced by the mass of the prism of soil within the trench and above the top of the sewer pipe. The second is the friction or shearing forces generated between the prism of soil in the trench and the sides of the trench.

The backfill soil has a tendency to settle in relation to the undisturbed soil in which the trench is excavated. This downward movement or tendency for movement induces upward shearing forces which support a part of the weight of the backfill. Thus, the resultant load on the horizontal plane at the top of the sewer pipe within the trench is equal to the weight of the backfill minus these upward shearing forces (Fig. 9-2).

Unusual conditions may be encountered in which poor natural soils may effect a change from trench to embankment conditions with considerably increased load on the sewer pipe. This is covered in the next subsection.

a. Use of Marston's Formula

Marston's formula for loads on rigid sewer pipe in trench conditions is:

$$W_c = C_d w B_d^2 \tag{9.2}$$

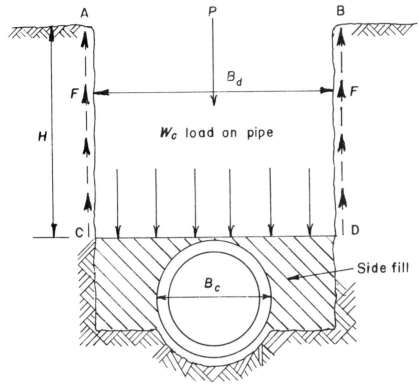

Fig. 9-2. Load-producing forces: P = weight of backfill ABCD; F = upward shearing forces on AC and BD; and, $W_c = P-2F$.

in which W_c is the load on the sewer pipe, in newtons per meter (pounds per foot); w is the density of backfill soil, in kilograms per cubic meter (pounds per cubic foot); B_d is the width of trench at the top of the sewer pipe, in meters (feet); C_d is a dimensionless load coefficient which is a function of the ratio of height of fill to width of trench and of the friction coefficient between the backfill and the sides of the trench. The load coefficient, C_d, is computed as follows:

$$C_d = \frac{1 - e^{-2\kappa\mu' \frac{H}{B_d}}}{2\kappa\mu'} \tag{9.3}$$

in which e is the base of natural logarithms and κ is Rankine's ratio of lateral pressure to vertical pressure:

$$\kappa = \frac{\sqrt{\mu^2 + 1} - \mu}{\sqrt{\mu^2 + 1} + \mu} = \frac{1 - \sin\phi}{1 + \sin\phi} \tag{9.4}$$

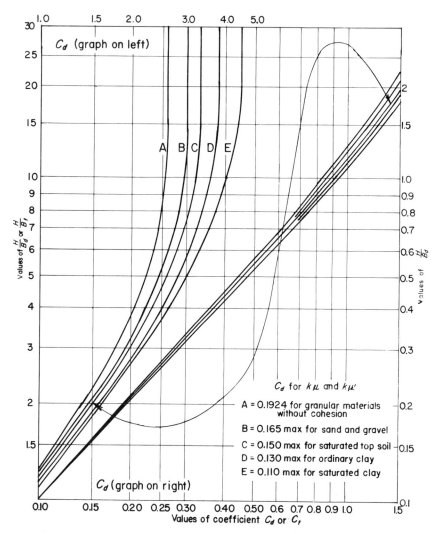

Fig. 9-3. Computation diagram for soil loads on trench installations (sewer pipe completely buried in trenches).

The other terms are: $\mu = \tan \phi =$ the coefficient of internal friction of backfill material; $\mu' = \tan \phi' =$ the coefficient of friction between backfill material and sides of trench (μ' may be equal to or less than μ, but never greater than μ); and H is the height of fill above top of pipe, in meters (feet). The value of C_d for various ratios of H/B_d and various types of soil backfill may be obtained from Fig. 9-3.

The trench load formula, Eq. 9.2, gives the total vertical load on a horizontal plane at the top of the sewer pipe. If the sewer pipe is rigid it will carry practically all this load. If the sewer pipe is flexible and the soil at the sides is compacted to the extent that it will deform under vertical load less than the sewer pipe itself will deform, the side fills may be expected to carry their proportional share of the total load. Under these circumstances the trench load formula may be modified to:

$$W_c = C_d w B_c B_d \tag{9.5}$$

in which B_c is the outside width of pipe, in meters (feet).

It is emphasized that Eq. 9.5 is applicable only if the backfill is compacted as described above. The equation should not be used merely because the pipe is a flexible type.

The term "side fill" refers to the soil backfill which is placed between the sides of sewer pipe and the sides of the trench. The character of this material and the manner of its placement have two important influences on the structural behavior of a sewer pipe.

First, the side fill may carry a part of the total vertical load on the horizontal plane at the elevation of the top of the sewer pipe.

Second, the side fill plays an important role in helping the sewer pipe carry vertical load. Every newton (pound) of force which can be brought to bear against the sides of an elastic ring increases the ability of the ring to carry vertical load by nearly the same amount.

Examination of Eq. 9.2 indicates the important influence the width of the trench exerts on the load as long as the trench condition formula applies. This influence has been verified by extensive experimental evidence. These experiments also have indicated that the width of trench at the top of the sewer pipe is the controlling factor.

The width of trench below the top of the sewer pipe also is important. It must not be permitted to exceed the safe limit for the strength of sewer pipe and class of bedding used. The minimum width must be consistent with the provision of sufficient working space at the sides of the sewer pipe to assemble joints properly, to insert and strip forms, and to compact backfill. The design engineer must allow reasonable tolerance in width for variations in field conditions and accepted construction practice.

For flexible pipe, narrow trench construction can result in some degree of soil arching, thereby reducing the load on the pipe. Conservative design practice for flexible pipe is based on the prism load. There are other approaches that may be considered for load calculation on flexible pipe. These approaches are discussed in a section below on "Flexible Sewer Pipe Design."

The position of the lower wale usually will determine the proper width of trench from face-to-face of sheeting, where sheeting and bracing are required. A working-room allowance of 300 mm (12 in.) from each side of the sewer pipe or sewer pipe cradle to the face of the sheeting is a workable minimum for small and medium-sized sewer pipe for trenches up to about 4 m (13 ft) deep.

At any given depth and for any given sewer pipe size there is a certain limiting value to the width of trench beyond which no additional load is transmitted to the sewer pipe. This limiting value is called the "transition width." There are sufficient experimental data at hand to show that it is safe to calculate the imposed load by means of the trench-conduit formula (Eq. 9.2) for all widths of trench less than that which gives a load equal to the load calculated by the projecting-conduit formula (discussed in the next section on "Embankment Conditions"). In other words, as the width of the trench increases (other factors remaining constant), the load on a rigid sewer pipe

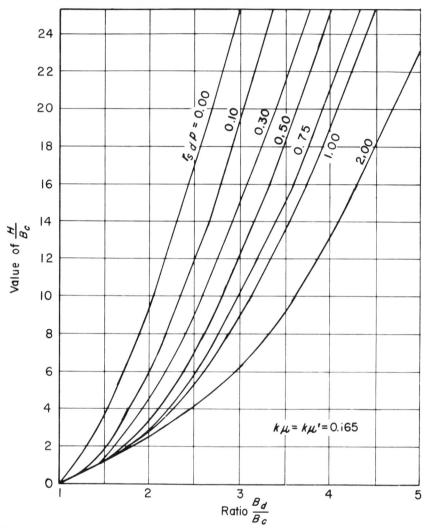

Fig. 9-4. Values of B_d/B_c at which pipe in trench and projecting pipe load formulas give equal loads.

increases in accordance with the theory for a trench sewer pipe until it equals the load determined by the theory for a projecting sewer pipe. The width of trench at which this transition occurs may be determined from Fig. 9-4 (3). The curves in Fig. 9-4 are calculated for sand and gravel, where $k\mu = 0.165$, but can be used for other types of soil since the change with varying values of $k\mu$ is small. In any event, the design engineer can check by calculating loads for both trench and embankment conditions. There is little research on the appropriate value of $r_{sd}p$ (the projection ratio times the settlement ratio) to use in the application of the transition width concept. In the absence of specific information, a value of $+0.5$ is suggested as a reasonably good working value. These quantities are defined in the subsection on "Loads for Embankment Conditions."

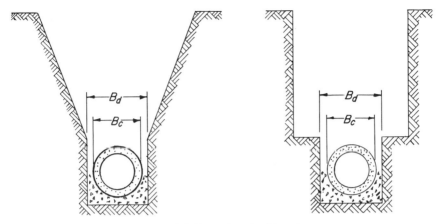

Fig. 9-5. Examples of subtrench.

It is advisable, in the structural design of sewers, to evaluate the effect of the transition width on both the design criteria and the construction latitude. A contractor, for instance, may wish to place well points for drainage in the trench. If this requires a wider trench than usual, a stronger sewer pipe or higher class of bedding may be necessary.

It may be economical and proper to excavate the trench with sloping sides in undeveloped areas where no inconvenience to the public or danger to property, buildings, subsurface structures, or pavements, will result. A subtrench (Fig. 9-5) may be used in such cases to minimize the load on the pipe. When sheeting of the subtrench at the pipe is necessary, it should extend about 450 mm (1.5 ft) above the top of the pipe.

When sheeting of the trench is necessary, it should be driven at least to the bottom of the pipe bedding or foundation material, if used. In general, in a constantly wet or dry area, sheeting and bracing should be left in place to prevent reduction in lateral support at the sides of the pipe because of voids formed by removal of the sheeting. Sheeting left in place should be cut off as far below the surface as practicable, but in no case less than 0.9 m (3 ft) below final ground elevation.

When wood sheeting is to be removed, the sheeting should be cut off 450 mm (1.5 ft) above the top of the pipe and the sheeting alongside the pipe left in place.

Steel sheeting to be removed should be pulled in increments as the trench is backfilled, and the soil should be compacted to prevent formation of voids. The portion of wood sheeting to be removed should be handled similarly.

Loads on sewer pipe in sheeted trenches should be calculated from a trench width measured to the outside of the sheeting if it is pulled or to the inside if it is left in place. Voids created by removal of the sheeting should be backfilled with a flowable material such as pea gravel.

If a shield is used in sewer pipe-laying operations, the shield width controls the width of the trench at the top of the sewer pipe. This width, with small addition for the space needed to advance the shield without a large friction loss, should be the width factor used in computing loads on the sewer pipe. Extreme care must be taken when advancing the shield in the trench to prevent the pipe joints from pulling apart and disturbance of the pipe bedding.

Sanitary sewers which are to be constructed in sloping-sided trenches with the slopes extending to the invert, or to any plane above the invert but below the top of the sewer, should be designed for loads computed by using the actual width of the trench at the top of the sewer pipe, or by the projecting-sewer formula (covered in the next subsection), whichever gives the least load on the sewer pipe.

If for any reason the trench becomes wider than that specified and for which the sewer pipe was designed, the load on the sewer pipe should be checked and a stronger sewer pipe or higher class of bedding used, if necessary.

b. Soil Characteristics-Trench Conditions

The load on a sewer pipe is influenced directly by the density of the soil backfill. This value varies widely for different soils, from a minimum of about 1,600 kg/m³ (100 lb/cu ft) to a maximum of about 2,200 kg/m³ (135 lb/cu ft). The average maximum unit weight of the soil which will constitute the backfill over the sewer pipe may be determined by density measurements in advance of the structural design of the sewer pipe. A design value of not less than 1,900 or 2,000 kg/m³ (120 or 125 lb/cu ft) is recommended if such measurements are not made.

The load also is influenced by the coefficient of friction between the backfill and the sides of the trench and by the coefficient of internal friction of the backfill soil. Ordinarily these two values will be nearly the same and may be so considered for design purposes, as in Fig. 9-3. However, in special cases this may not be true. For example, if the backfill is sharp sand and the sides of the trench are sheeted with finished lumber, μ may be substantially greater than μ'. Unless specific information to the contrary is available, values of the products $k\mu$ and $k\mu'$ may be assumed to be the same and equal to 0.130. If the backfill soil is a "slippery" clay and there is a possibility that it will become very wet shortly after being placed, $k\mu$ and $k\mu'$ equal to 0.110 (maximum for saturated clay, Fig. 9-3) should be used.

Sample Calculations:

Example 9-1. Determine the load on a 24-in. diameter rigid sewer pipe under 14 ft of cover in trench conditions.

Assume that the sewer pipe wall thickness is 2 in.; $B_c = 24 + 4 = 28$ in. $= 2.33$ ft; $B_d = 2.33 + 2.00 = 4.33$ ft; and $w = 120$ lb/cu ft for saturated top soil backfill. Then $H/B_d = 14/4.33 = 3.24$; C_d (from Fig. 9-3) $= 2.1$; and $W_c = 2.1 \times 120 \times (4.33)^2 = 4,720$ lb/ft (68,880 N/m).

Example 9-2. Determine the load on the same size sewer laid on a concrete cradle and with trench sheeting to be removed.

Assume that the wall thickness is 2 in.; the cradle projection outside of the sewer pipe is 8 in. (4 in. on each side); and the maximum clearance between cradle and outside of sheeting is 14 in. Then $B_d = 24 + (2 \times 2 \text{ in.}) + 8 + (2 \times 14) = 64$ in. $= 5.33$ ft.

As this seems to be an extremely wide trench, a check should be made on the transition width of the trench: $B_c = 2.33$ ft; $H = 14$ ft; $r_{sd}p = 0.5$; and $H/B_c = 14.2.33 = 6.0$.

From Fig. 9-4, $B_d/B_c = 2.39$ (the ratio of the width of the trench to the width of the sewer at which loads are equal by both ditch-sewer theory and projecting-sewer theory); $B_d = 2.33 \times 2.39 = 5.57 > 5.33$; $H/B_d = 14/5.33 = 2.63$; C_d (from Fig. 9-3) $= 1.85$; and $W_c = 1.85 \times 120 \times (5.33)^2 = 6,300$ lb/ft (91,700 N/m).

Example 9-3. Determine the load on the same sewer if (rough) sheeting is

left in place.
B_d becomes 4 in. less = 5.00 ft; H/B_d = 14/5.00 = 2.8; C_d (from Fig. 9-3) = 1.92; and W_c = 1.92 × 120 × (5.00)² = 5,750 lb/ft (84,040 N/m).

Example 9-4. Determine the load on a 30-in. diameter flexible sewer pipe installed in a trench 4 ft 6 in. wide at a depth of 12 ft.

Assume the soil is clay weighing 120 lb/cu ft. and that it will be well compacted at the sides of the sewer pipe. Then H = 12 ft; B_d = 4.5 ft; B_c = 2.5 ft; H/B_d = 2.67; C_d = 1.9; and W_c = 1.9 × 120 × 4.5 × 2.5 = 2,565 lb/ft (37,450 N/m).

For conservative design, the prism load should be determined. The prism load on flexible sewer pipe will be W = 2.5 × 12 × 120 = 3,600 lb/ft (52,460 N/m).

4. Loads for Embankment Conditions

A sewer pipe is described as a projecting sewer pipe when it is installed in a wide trench or in such a manner that the top of the sewer pipe is at or near the natural ground surface or the surface of thoroughly compacted soil and subsequently is covered with an embankment. If the top of the sewer pipe projects some distance above the natural ground surface or if it is installed in a wide trench, it is a positive projecting sewer pipe. There are, however, other methods of installing sewer pipe under embankments which have the favorable effect of minimizing the load on the sewer pipe. In these cases, the installation is classified as a negative projecting sewer pipe or an induced trench sewer pipe (Fig. 9-1).

These variations of embankment conditions will be treated separately for convenience in computation.

a. Positive Projecting Sewer Pipe

The load on a positive projecting sewer pipe is equal to the weight of the prism of soil directly above the structure, plus (or minus) vertical shearing forces which act on vertical planes extending upward into the embankment from the sides of the sewer pipe. For an embankment installation of sufficient height, these vertical shearing forces may not extend to the top of the embankment, but terminate in a horizontal plane at some elevation above the top of the sewer pipe known as the "plane of equal settlement" as shown in Fig. 9-6. The shear increment acts downward when $(s_m + s_g) > (s_f + d_c)$ and vice versa. In this expression s_m is the compression of the columns of soil of height pB_c; s_g is the settlement of the natural ground adjacent to the sewer pipe; s_f is the settlement of the bottom of the sewer pipe; and d_c is the deflection of the sewer pipe.

The location of the plane of equal settlement is determined by equating the total strain in the soil above the pipe to that in the side fill plus the settlement of the critical plane. When the plane of equal settlement is an imaginary plane above the top of the embankment, i.e., shear forces extend to the top of the embankment, the installation is called either "complete trench condition" or "complete projection condition," depending on the direction of the shear forces. When the plane of equal settlement is located within the embankment (Fig. 9-6), the installation is called "incomplete trench condition," or "incomplete projection condition" (Fig. 9-7).

In computing the settlement values, the effect of differential settlement caused by any compressible layers below the natural ground surface also must be considered. An exceptional situation for a sewer pipe in a trench can be encountered where the natural soil settles more than the trench backfill, such

GRAVITY SANITARY SEWER

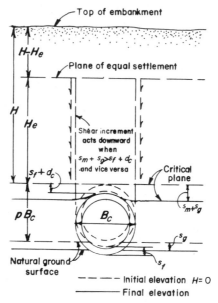

Fig. 9-6. Settlements that influence loads on positive projecting sewer pipe: s_g = settlement of natural ground adjacent to sewer pipe, s_m = compression of columns of soil of height pB_c, d_c = deflection of sewer pipe, and s_f = settlement of bottom of sewer pipe.

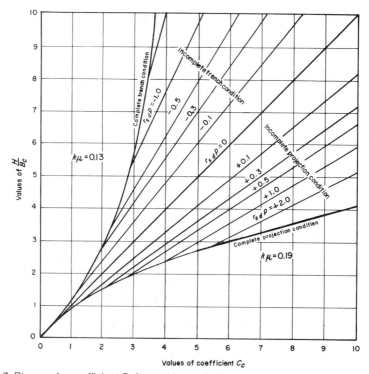

Fig. 9-7. Diagram for coefficient C_c for positive projecting sewer pipes.

as where the natural soils are organic or peat, and the trench backfill is relatively incompressible compacted fill. A more common situation is where the sewer pipe is pile-supported in organic soils. In such cases, the load on the sewer pipe is greater than that of the prism above the pipe, and down drag loads should be considered in the design of the piles.

Marston's Formula
Marston's formula for loads on rigid positive projecting sewer pipe is written:

$$W_c = C_c w B_c^2 \tag{9.6}$$

in which W_c is the load on the sewer pipe, in newtons/meter (pounds per foot); B_c is the outside width of the sewer pipe, in meters (feet); and C_c is the load coefficient. Values of C_c may be obtained from Fig. 9-7. In this diagram, H is the height of fill above the top of the sewer pipe, in meters (feet); B_c is the outside width of sewer pipe, in meters (feet); p is the projection ratio; and r_{sd} is the settlement ratio (the latter two terms are defined in the next subsection).

Influence of Environmental Factors
The shear component of the total load on a sewer pipe under an embankment depends on two factors associated with the conditions under which the sewer pipe is installed. These are the projection ratio and the settlement ratio.

The projection ratio, p, is defined as the ratio of the distance that the top of the sewer pipe projects above the adjacent natural ground surface, or the top of thoroughly compacted fill, or the bottom of a wide trench, to the vertical outside height of the sewer pipe. It is a physical factor that can be determined in advanced stages of planning when the size of the sewer pipe and its elevation have been established.

The settlement ratio, r_{sd}, indicates the direction and magnitude of the relative settlements of the prism of soil directly above the sewer pipe and of the prisms of soil adjacent to it.

In computing the settlement, the influence of any compressible layers below the sewer pipe also must be considered.

These relative settlements generate the shearing forces which combine algebraically with the weight of the central prism of soil to produce the resultant load on the sewer pipe. The settlement ratio is the quotient obtained by taking the difference between the settlement of the horizontal plane in the adjacent soil which was originally level with the top of the sewer pipe (the critical plane) and the settlement of the top of the sewer pipe, and dividing the difference by the compression of the columns of soil between the natural ground surface and the level of the top of the sewer pipe. The formula for the settlement ratio is:

$$r_{sd} = \frac{(s_m + s_g) - (s_f + d_c)}{s_m} \tag{9.7}$$

in which r_{sd} is the settlement ratio; s_g is the settlement of the natural ground adjacent to the sewer pipe; s_m is the compression of the columns of soil of height pB_c; $(s_m + s_g)$ is the settlement of the critical plane; d_c is the deflection of

Table 9-1 Recommended Design Values of r_{sd}.

Type of Sewer Pipe (1)	Soil Conditions (2)	Settlement Ratio, r_{sd} (3)
Rigid	Rock or unyielding foundation	+1.0
Rigid	Ordinary foundation	+0.5 to +0.8
Rigid	Yielding foundation	0 to +0.5
Rigid	Negative projecting installation	−0.3 to −0.5
Flexible	Poorly compacted side fills	−0.4 to 0
Flexible	Well compacted side fills	0

the sewer pipe, that is, the shortening of its vertical dimension; s_f is the settlement of the bottom of the sewer pipe; and $(s_f + d_c)$ is the settlement of the top of the sewer pipe.

The elements of the settlement ratio are shown in Fig. 9-6. When the settlement ratio is positive, the shearing forces induced along the sides of the central prism of soil are directed downward, and the load on the sewer pipe is greater than the weight of the central prism. When the settlement ratio is negative, the shearing forces act upward and the load is less than the weight of the central prism.

The numerical magnitude of the product of the projection ratio and the settlement ratio, $r_{sd}\,p$, is an indicator of the relative height of the plane of equal settlement and, therefore, of the magnitude of the shear component of the load. The plane of equal settlement is at the top of the sewer pipe when this product is equal to zero. There are no induced shearing forces in this case, and the load is equal to the weight of the central prism (the "prism load").

It is not practical to predetermine a value of the settlement ratio by estimating the magnitude of its various elements except in very general terms. Rather, it should be treated as an empirical factor. Recommended design values of r_{sd}, based on measured settlements of a number of actual installations, are given in the Table 9-1.

The last three cases (Table 9-1) presume soil conditions immediately under the sewer pipe to be the same as those in the adjacent areas outside the trench. In these cases, the settlement ratio may be conservatively assumed as zero in locations with highly fluctuating water tables above the pipe or plastic native trench soils. This results in designing for the "prism load," or the weight of the prism of soil above the pipe. In such cases, C_c is equal to H/B_c and Marston's formula for the prism load becomes:

$$W_c = HwB_c \tag{9.8}$$

The prism load may also be expressed in terms of soil pressure, P, in pascals (pounds per square foot) at depth H as:

$$P = wH = \frac{W_c}{B_c} \tag{9.9}$$

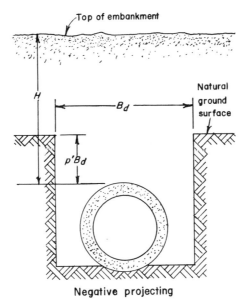

Fig. 9-8. Negative projecting sewer pipe.

Embankment Soil Characteristics

The load on a projecting sewer pipe is influenced directly by the density of the embankment soil. If the soil is to be compacted to a specified dry density, the corresponding wet density under normal moisture conditions should be used in calculating the load. A design value of not less than 1,900 or 2,000 kg/m³ (120 or 125 lb/cu ft) is recommended if specific information relative to soil density is not available.

The load also is influenced by the coefficient of internal friction of the embankment soil. Recommended values of the product $k\mu$ (Fig. 9-7) are:

for a positive settlement ratio, $k\mu = 0.19$;
for a negative settlement ratio, $k\mu = 0.13$.

b. Negative Projecting and Induced Trench Sewer Pipes

A negative projecting sewer pipe (Fig. 9-8) is one installed in a relatively shallow trench with its top at some elevation below the natural ground surface. The trench above the sewer pipe is refilled with loose, compressible material, and the embankment is constructed to finished grade by ordinary methods.

Sometimes straw, hay, cornstalks, sawdust, or similar materials may be added to the trench backfill to augment the settlement of the interior prism. The greater the value of the negative projection ratio, p', and the more compressible the trench backfill over the sewer pipe, the greater will be the settlement of the interior prism of soil in relation to the adjacent fill material. In using this technique, the plane of equal settlement must fall below the top of the finished embankment. This action generates upward shearing forces which relieve the load on the sewer pipe.

An induced trench sewer pipe (Fig. 9-9) first is installed as a positive projecting sewer pipe. The embankment then is built up to some height above the top and thoroughly compacted as it is placed. A trench of the same width

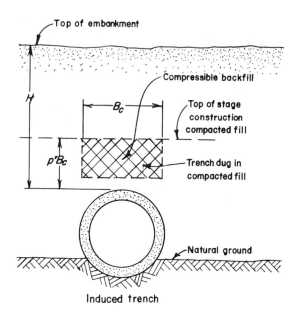

Fig. 9-9. Induced trench pipe.

as the sewer pipe next is excavated directly over the sewer pipe down to or near its top. This trench is refilled with loose, compressible material, and the balance of the embankment is completed in a normal manner.

The formula for loads on negative projecting sewer pipe is:

$$W_c = C_n w B_d^2 \qquad (9.10)$$

in which W_c is the load on the sewer pipe, in newtons per meter (pounds per foot); w is the density of soil, in kilograms per cubic meter (pounds per cubic foot); B_d is the width of the trench, in meters (feet); C_n is the load coefficient (Fig. 9-10), a function of H/B_d or H/B_c, p', and r_{sd}; p' is the projection ratio; and r_{sd} is the settlement ratio as defined below.

In the case of the induced trench sewer pipe, B_c is substituted for B_d in Eq. 9-10, in which B_c is the width of the sewer pipe in meters (feet), assuming the trench in the fill is no wider than the sewer pipe.

The projection ratio, p', is equal to the vertical distance from the firm ground surface down to the top of the sewer pipe divided by the width of the trench, B_d, in the case of negative projecting sewer pipe, or by the width of the sewer pipe, B_c, in the case of induced trench sewer pipe.

The settlement ratio, r_{sd}, for these cases is the quotient obtained by taking the difference between the settlement of the firm ground surface and the settlement of the plane in the trench backfill which was originally level with the ground surface (the critical plane) and dividing the difference by the compression of the column of soil in the trench. The formula for the settlement ratio is:

$$r_{sd} = \frac{s_g - (s_d + s_f + d_c)}{s_d} \qquad (9.11)$$

STRUCTURAL REQUIREMENTS

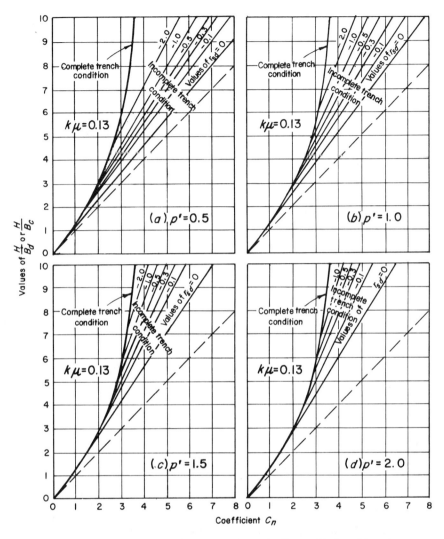

Fig. 9-10. Diagrams for coefficient C_n for negative projecting and induced trench sewer pipes.

in which r_{sd} is the settlement ratio for negative projecting or induced trench sewer pipe; s_g is the settlement of the firm ground surface; s_d is the compression of trench backfill within the height $p'B_d$ or $p'B_c$; s_f is the settlement of the bottom of the sewer pipe; d_c is the deflection of the sewer pipe, that is, the shortening of its vertical dimension; and $(s_d + s_f + d_c)$ is the settlement of the critical plane. The elements of the settlement ratio are shown in Fig. 9-11. Present knowledge of the value of the settlement ratio for induced trench sewer pipe is meager. In the absence of extensive factual data relative to probable values of the settlement ratio, it is tentatively recommended that this ratio be assumed to lie between -0.3 and -0.5. Recent research reported by Taylor (29) of the Illinois Division of Highways indicated that the measured

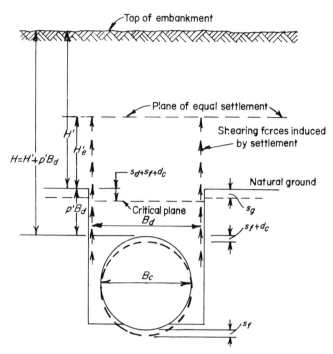

Fig. 9-11. Settlements that influence loads on negative projecting sewer pipes.

settlement ratio of 48-in. (1,200 mm) reinforced concrete pipe culvert, installed as an induced trench conduit under 9 m (30 ft) of fill, varied from −0.25 to −0.45.

c. Sewer Pipe Under Sloping Embankment Surfaces

Cases arise when the sewer pipe has different heights of fill on the two sides because of the sloping surface of the embankment or when embankment exists on one side of the sewer pipe only. Such cases require special analysis. Design based on the larger fill height may not yield conservative results. Where yielding ground may envelope the sewer pipe, a surcharge on one side of the sewer pipe may result in vertical displacement.

Sample Calculations:

Example 9-5. Determine the load on a 48-in. diameter reinforced concrete sewer pipe installed as a positive projecting pipe under a fill 32 ft high above the top of the pipe. The wall thickness of the sewer pipe is 5 in. and the density of the fill is 125 lb/cu ft.

Assume the projection ratio is +0.5 and the settlement ratio is +0.6. Then $H = 32$ ft; $B_c = 4.83$ ft; $H/B_c = 6.63$; $r_{sd}p = 0.5 \times 0.6 = 0.3$; C_c (from Fig. 9-7) $= 9.2$; and $W_c = 9.2 \times 125 \times (4.83)^2 = 26,800$ lb/ft (392,300 N/m).

Example 9-6. Determine the load on the sewer pipe of *Example 9-5* when installed as a negative projecting sewer pipe in a trench whose depth is such that the top of the sewer pipe is 7 ft below the surface of the natural ground in which the trench is dug. The width of the trench is 2 ft greater than the outside diameter of the sewer pipe.

Assume the settlement ratio = -0.3. Then $H = 32$ ft; $B_d = 4.83 + 2 = 6.83$ ft; $H/B_d = 4.69$; $p' = 1.0$; C_n (from Fig. 9-10) = 3.0; and $W_c = 3.0 \times 125 \times (6.83)^2 = 17{,}500$ lb/ft (255,950 N/m).

Example 9-7. Determine the load on the sewer pipe of *Example 9-5* when installed as an induced trench sewer pipe with its top 2.5 ft below the elevation to which the soil is compacted thoroughly for a distance of 12 ft on each side of the sewer pipe.

Assume the settlement ratio = -0.3. Then $H = 32$ ft; $B_c = 4.83$; $H/B_c = 6.63$; $p' =$ approximately 0.5; C_n (from Fig. 9-10) = 4.8; and $W_c = 4.8 \times 125 \times (4.83)^2 = 14{,}000$ lb/ft (204,300 N/m).

5. Loads for Jacked Sewer Pipe and Certain Tunnel Conditions

When the sanitary sewer is more than 9 to 12 m (30 to 40 ft) deep or when surface obstructions are such that it is difficult to construct the sanitary sewer by trenching, it may be more economical to place the sanitary sewer by means of jacking or tunneling. The theories set forth in this Manual usually will be appropriate for materials where jacking of the sewer pipe is possible and for tunnels in homogeneous soils of low plasticity. Where a tunnel is to be constructed through materials subject to unusually high internal pressures and stresses, such as some types of clays or shales which tend to squeeze or swell, or through blocky and seamy rock, the loads on the sewer pipe cannot be determined from the factors discussed here. Reference should be made to the following section on tunnels.

The methods of constructing sanitary sewers by tunneling and jacking are described in Chapter 10. Tunnel supports carry the earth load until the sewer pipe is constructed and the voids between the sewer pipe and tunnel supports are filled. Jacked sewer pipe (4,5) is assumed to carry the earth load as it is pushed into place.

a. Load-Producing Forces

For the materials considered in this Manual, the vertical load acting on the jacked sewer pipe or tunnel supports, and eventually the sewer pipe in the tunnel, is the resultant of two major forces. First is the weight of the overhead prism of soil within the width of the jacked sewer pipe or tunnel excavation. Second is the shearing forces generated between the interior prisms and the adjacent material caused by the internal friction and cohesion of the soil.

During excavation of a tunnel, and varying somewhat with construction methods, the soil directly above the face of the tunnel tends to settle slightly in relation to the soil adjacent to the tunnel because of the lack of support during the period immediately after excavation and prior to placement of the tunnel support. Also, the tunnel supports and the sewer pipe must deflect and settle slightly when the vertical load comes on them. This downward movement or tendency for movement induces upward shearing forces which support a part of the weight of the prism of earth above the tunnel. In addition, the cohesion of the material provides further support for the weight of the prism of earth above the tunnel. The resultant load on the horizontal plane on the top of the tunnel and within the width of the tunnel excavation is equal to the weight of the prism of earth above the tunnel minus the upward friction forces and cohesion of the soil along the limits of the prism of soil over the tunnel.

Hence, the forces involved with gravity earth loads on jacked sewer pipe or tunnels in such soils are similar to those discussed for loads on sewer pipe in trenches except for the cohesion of the material. Cohesion also exists in the case of loads in trenches and embankments but is neglected because the

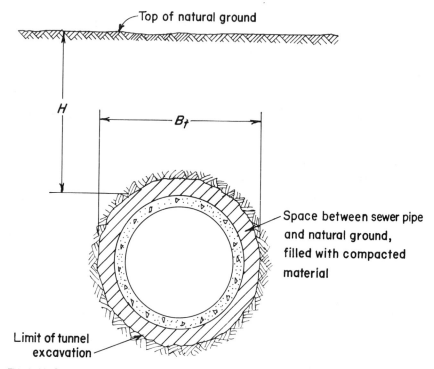

Fig. 9-12. Sewer pipe in tunnel.

cohesion of the disturbed soil is of minor consequence and may be absent altogether if the soil is saturated. However, in the case of jacked sewer pipe or in tunnels, where the soil is undisturbed, cohesion can be an appreciable factor in the loads and may be considered safely if reasonable coefficients are assumed.

Jacking stresses must be investigated in pipe which is to be jacked into place. The critical section is at the pipe joint where the transfer of stress from one pipe to the adjacent pipe occurs. Jointing materials should be used which will provide uniform bearing around the pipe circumference (5). Thrust at the joint is usually transmitted through the tongue or groove but not both. Concrete stress in the tongue or groove should be checked and additional reinforcement for both longitudinal and bursting stresses provided if required.

b. Marston's Formula

When modified to include cohesion, Marston's formula may be used to determine the gravity soil loads on jacked sewer pipe or sewer pipe in tunnels through undistrubed soil (Fig. 9-12). It takes the form:

$$W_t = C_t B_t (w B_t - 2c) \tag{9.12}$$

in which W_t is the load on the sewer pipe or tunnel support, in newtons per meter (pounds per foot); w is the density of the soil above the tunnel, in kilograms per cubic meter (pounds per cubic foot); B_t is the maximum width of

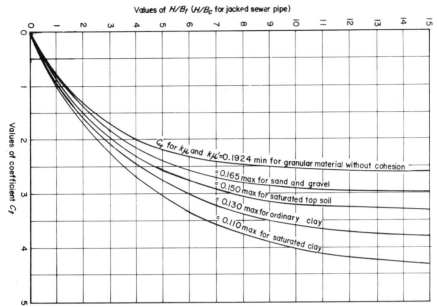

Fig. 9-13. Diagram for coefficient C_t for jacked sewer pipe or tunnels in undisturbed soil.

the tunnel excavation, in meters (feet) (B_c in the case of jacked sewer pipe); c is the cohesion coefficient, in pascals (pounds per square foot); and C_t is a load coefficient which is a function of the ratio of the distance from the ground surface to the top of the tunnel, to the width of the tunnel excavation and of the coefficient of internal friction of the natural material above the tunnel.

The formula for C_t is identical to that for C_d (Eq. 9-3), except that H is the distance from the ground surface to the top of the tunnel and B_t is substituted for B_d. The values of the coefficient for C_t for various ratios of H/B_t and various types of materials may be obtained from Fig. 9-13 or Fig. 9-3. Values of $k\mu$ and $k\mu'$ are the same as those noted in Fig. 9-3.

An analysis of the formula for computing C_t indicates that for very high values of H/B_t the coefficient C_t approaches the limiting value of $1/(2k\mu')$. Hence, where the tunnel is very deep, the load on the tunnel can be calculated readily by using the limiting value of C_t.

c. Tunnel Soil Characteristics

The discussion regarding unit weight and coefficient of friction for sanitary sewers in trenches applies equally to the determination of earth loads on jacked sewer pipe or sewer pipe in tunnels through undisturbed soil.

The one additional factor that enters into the determination of loads on tunnels is c, the coefficient of cohesion. An examination of Eq. 9.12 shows that the proper selection of c is very important; unfortunately, it can vary widely even for similar types of soils.

It may be possible in some instances to obtain undisturbed samples of the material and to determine the value of c by appropriate laboratory tests. Such testing should be done whenever possible. It is suggested that conservative values of c be used to allow for a saturated condition of the soil or for other unknown factors. Design values should probably be about 33% of the

Table 9-2 Recommended Safe Values of Cohesion, c

Material (1)	Values of c	
	in kilopascals (2)	in pounds per square foot (3)
Clay, very soft	2	40
Clay, medium	12	250
Clay, hard	50	1,000
Sand, loose dry	0	0
Sand, silty	5	100
Sand, dense	15	300

laboratory test value to allow for uncertainties.

Recommended safe values of cohesion for various soils (if it is not practicable to determine c from laboratory tests) are shown in Table 9-2.

It is suggested that the value of c be taken as zero in the zone subject to seasonal frost and cracking because of dessication or loss of strength from saturation. In addition, a minimum value for $(wB_t - 2c)$ should be assumed in cases where $2c$ approaches wB_t.

d. Effect of Excessive Excavation

Where the tunnel is constructed by a method that results in excessive excavation and where the voids above the sewer pipe or tunnel lining are not backfilled carefully, or packed with grout or other suitable backfill materials, saturation of the soil or vibration eventually may destroy the cohesion of the undisturbed material above the sewer pipe and result in loads in excess of those calculated using Eq. 9.12. If this situation is anticipated, it is suggested that Eq. 9.12 be modified by eliminating the cohesion term. The calculated loads then will be the same as those obtained from Eq. 9.2.

Sample Calculation:

Example 9-8. Assume the width of excavation, $B_t = 78$ in. $= 6.5$ ft; type of soil is silty sand ($k\mu' = 0.150$, $c = 100$ psf, and $w = 110$ lb/cu ft); and the depth of tunnel, $H = 40$ ft. Then $H/B_t = 40/6.5 = 6.15$; C_t (from Fig. 9-13) $= 2.83$.

Employing Eq. 9.12, $W_t = 2.83 \times 6.5 (110 \times 6.5 - 2 \times 100)$; or $W_t = 9{,}500$ lb/ft (138,300 N/m).

If the tunnel were very deep, $C_t = 1/(2k\mu') = 3.33$; and $W_t = 11{,}200$ lb/ft (162,700 N/m).

6. Loads for Tunnels

When the sanitary sewer is to be constructed in a tunnel through homogeneous soils of low plasticity, design should be based on the theories set forth in the previous section describing jacked sewer pipe. The design of tunnels through other types of materials is discussed in this section. The usual procedure in tunnel construction is to complete the excavation first and then place either a cast-in-place concrete liner or sewer pipe then grouting or concreting it in place. Additional strength in such a section can be obtained by means of pressure grouting to strengthen the surrounding material instead of relying totally on the liner or pipe itself. Tunnel loads, therefore, usually are determined for purposes of selecting supports to be used during excavation, and the sewer pipe or cast-in-place liner is designed primarily to withstand loads from pressure grouting.

A complete discussion of tunnels is not within the scope of this manual, and the design engineer's attention is called to references listed at the end of this chapter (6–9).

a. Load-Producing Forces

When the tunnel is to be constructed through soils which tend to squeeze or swell, such as some types of clay or shale, or through blocky to seamy rock, the vertical load cannot be determined from a consideration of the factors discussed previously, and Eq. 9.12 is not applicable.

The determination of rock pressures exerted against the tunnel lining is largely an estimate based on previous experience of the performance of linings in similar rock formations, although attempts at numerical analysis of stress conditions around a tunnel shaft have been made.

In the case of plastic clay, the full weight of the overburden is likely to come to rest on the tunnel lining some time after construction. The extent of lateral pressures to be expected has not as yet been determined fully, especially the passive resistance which will be maintained permanently by a plastic clay in the case of a flexible ring-shaped tunnel lining. For normally consolidated clays, suggested lateral pressures are on the order of 2/3 to 7/8 of vertical overburden pressures.

On the other hand, when tunneling through sand, only part of the weight of the overburden will come to rest on the tunnel lining at any time if adequate precautions are taken. The relief will be the result of the transfer of the soil weight immediately above the tunnel to the adjoining soil mass by shearing stresses along the vertical planes. In this case Marston's formula may be used for estimating the total load which the tunnel lining may have to carry.

Great care must be taken to prevent any escape of sand into the tunnel during its construction. Moist sand usually will arch over small openings and not cause trouble in this respect; however, entirely dry sand, which is sometimes encountered, is liable to trickle into the tunnel through gaps in the temporary lining. Wet sand or sand under the natural water table will flow readily through the smallest gaps. Sand movements of this kind destroy most if not all of the arching around the tunnel, with a resulting strong increase of both vertical and horizontal pressures on the supports of the lining. Such cases have been recorded and have caused considerable difficulty.

All soil parameters required for design should be obtained from laboratory testing.

7. Alternate Design Method

For large diameter sanitary sewers, such as greater than 1,200 mm (48-in.) in diameter, designs based on the Marston method may yield conservative results. In such cases a more precise analysis can be made utilizing the principles of soil-structure interaction. Analysis should consider both the geometry of the system and material properties of the sewer pipe and the surrounding soil mass.

A simpler method based on arch analysis, which considers only the geometry of the sewer pipe and section properties of the sewer pipe material (16), can also be used for any specified loading condition.

In the method of soil-structure interaction analysis, loadings on the

Table 9-3 "At-Rest" Pressure Coefficients

Soil Type (1)	"At-Rest" Coefficient (2)
Granular soils	0.5 to 0.67
Cohesive soils, medium to hard	0.67 to 0.88
Cohesive soils, soft	0.75 to 1.0

sewer pipe are automatically generated from the specified boundary conditions, the material properties, and the constitutive relationships of material behavior. Most solutions consider elastic behavior of the materials. Elastoplastic behavior and nonlinear analysis are also available.

The arch analysis method requires specification of vertical and lateral loads. The vertical loads can be determined by the Marston method, as described in preceding sections, and distributed uniformly over the full width of the sewer pipe. Lateral loads depend on the soil type and geologic history of the soil deposit. Design parameters should be obtained from a soils consultant knowledgeable of the subsurface conditions in the area. For sewer pipe installed in tunnel or in a trench with properly compacted backfill, the recommended design lateral pressures are those corresponding to "at-rest" conditions. Where the backfill on the sides of the sanitary sewer may be loosely placed or insufficiently compacted, "active" pressure coefficients should be used to determine the lateral pressures. For preliminary analysis, the "at-rest" pressure coefficients in Table 9-3 are suggested. Since active and passive earth pressures are the result of lateral strain in the soil mass, the at-rest condition refers to the lateral pressures existing in a large soil mass not subject to horizontal forces or strains except those resulting from its own weight.

C. SUPERIMPOSED LOADS ON SANITARY SEWERS

1. General Method

Two types of superimposed loads are encountered commonly in the structural design of sanitary sewers, concentrated load and distributed load. Loads on sewer pipe caused by these loadings can be determined by application of Boussinesq's solution for stresses in a semi-infinite elastic medium through the convenience of an integration developed by D.L. Holl for concentrated loads and tables of influence coefficients developed by Newmark for distributed loads (26).

Other methods, such as that given in the AASHTO Code, can be used to determine loads on sewer pipe from superimposed loads (27). The AASHTO method is intended for use with wheel loads directly over the pipe and may not be conservative or applicable for other types of loads, such as those from adjacent building foundations. Empirical studies indicate the difficulties of accurately predicting the actual loads on the pipe. Therefore, the method presented in this text is based on the more general and theoretically correct Boussinesq equations.

In the design of buried sewer pipe systems, proper consideration of construction loads is necessary. Construction loads resulting from heavy equipment and reduced backfill heights can produce loads on the sewer pipe that exceed final design loads.

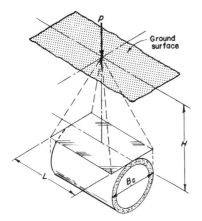

Fig. 9-14. Concentrated superimposed load vertically centered over sewer pipe.

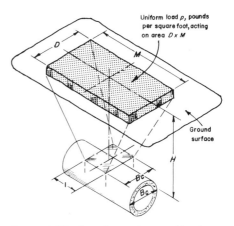

Fig. 9-15. Distributed superimposed load vertically centered over sewer pipe (psf × 47.9 = Pa).

2. Concentrated Loads

The formula for load caused by a superimposed concentrated load, such as a truck wheel (Fig. 9-14), is given the following form by D.L. Holl's integration of Boussinesq's formula:

$$W_{sc} = C_s \frac{PF}{L} \qquad (9.13)$$

in which W_{sc} is the load on the sewer pipe, in newtons per unit length (pounds per unit length); P is the concentrated load, in newtons (pounds); F is the impact factor; C_s is the load coefficient (Table 9-4), a function of $B_c/(2H)$ and $L/(2H)$; H is the height of fill from the top of sewer pipe to ground surface, in meters (feet); B_c is the width of sewer pipe, meters (feet); and L is the effective length of sewer pipe, in meters (feet).

The effective length of a sewer pipe is defined as the length over which the average load caused by surface traffic wheels produces nearly the same stress in the sewer pipe wall as does the actual load which varies in intensity from point to point. Little research information is available on this subject. Tentative recommendations are to use an effective length equal to 1.0 m (3 ft) for sewer pipe greater than 1.0 m (3 ft) long. The actual length should be used for sewer pipe shorter than 1.0 m (3 ft).

If the concentrated load is displaced laterally and longitudinally from a vertically centered location over the section of sewer pipe under construction, the load on the pipe can be computed by adding algebraically the effect of the concentrated load on various rectangles each with a corner centered under the concentrated load. Values of C_s in Table 9-4 divided by 4 equal the load coefficient for a rectangle whose corner is vertically centered under the concentrated load.

3. Impact Factor

The impact factor, F, reflects the influence of dynamic loads caused by

Table 9-4 Values of Load Coefficients, C_s, for Concentrated and Distributed Superimposed Loads Vertically Centered over Sewer Pipe[a]

$\frac{D}{2H}$ or $\frac{B_c}{2H}$	\multicolumn{14}{c}{$\frac{M}{2H}$ or $\frac{L}{2H}$}													
	0.1	0.2	0.3	0.4	0.5	0.6	0.7	0.8	0.9	1.0	1.2	1.5	2.0	5.0
(1)	(2)	(3)	(4)	(5)	(6)	(7)	(8)	(9)	(10)	(11)	(12)	(13)	(14)	(15)
0.1	0.019	0.037	0.053	0.067	0.079	0.089	0.097	0.103	0.108	0.112	0.117	0.121	0.124	0.128
0.2	0.037	0.072	0.103	0.131	0.155	0.174	0.189	0.202	0.211	0.219	0.229	0.238	0.244	0.248
0.3	0.053	0.103	0.149	0.190	0.224	0.252	0.274	0.292	0.306	0.318	0.333	0.345	0.355	0.360
0.4	0.067	0.131	0.190	0.241	0.284	0.320	0.349	0.373	0.391	0.405	0.425	0.440	0.454	0.460
0.5	0.079	0.155	0.224	0.284	0.336	0.379	0.414	0.441	0.463	0.481	0.505	0.525	0.540	0.548
0.6	0.089	0.174	0.252	0.320	0.379	0.428	0.467	0.499	0.524	0.544	0.572	0.596	0.613	0.624
0.7	0.097	0.189	0.274	0.349	0.414	0.467	0.511	0.546	0.584	0.597	0.628	0.650	0.674	0.688
0.8	0.103	0.202	0.292	0.373	0.441	0.499	0.546	0.584	0.615	0.639	0.674	0.703	0.725	0.740
0.9	0.108	0.211	0.306	0.391	0.463	0.524	0.574	0.615	0.647	0.673	0.711	0.742	0.766	0.784
1.0	0.112	0.219	0.318	0.405	0.481	0.544	0.597	0.639	0.673	0.701	0.740	0.774	0.800	0.816
1.2	0.117	0.229	0.333	0.425	0.505	0.572	0.628	0.674	0.711	0.740	0.783	0.820	0.849	0.868
1.5	0.121	0.238	0.345	0.440	0.525	0.596	0.650	0.703	0.742	0.774	0.820	0.861	0.894	0.916
2.0	0.124	0.244	0.355	0.454	0.540	0.613	0.674	0.725	0.766	0.800	0.849	0.894	0.930	0.956

[a]Influence coefficients for solution of Holl's and Newmark's integration of the Boussinesq equation for vertical stress.

Table 9-5 Suggested Values of Impact Factor, F

Traffic Type (1)	F (2)
Highway	1.30
Railway	1.40
Airfield runways (for taxiways, consult FAA)	1.00

traffic at the ground surface. Suggested values for various kinds of traffic are shown in Table 9-5.

The impact effect decreases with increasing cover. The AASHTO (highway) Code (27) recommends a reduction to 1.00 where depth of cover exceeds 1 m (3 ft) or the pipe outside diameter, whichever is larger. The AREA (railway) Code (30) recommends 10 ft (3 m) of cover for the elimination of impact effect. In design of airfield pavements, it is customary not to design for impact on runways because of the counterbalancing effect of the lift provided by aircraft wings. Similarly, for taxiways the slower speed reduces the lift, but it also is considered to reduce impact to a negligible amount in most cases. Since airfield pavement design involves empirical procedures, the design engineer should exercise judgment as to the amount of impact to be included in the design of buried sewer pipes. Common practice is to use an impact factor of 1.0 for runways and 1.5 for taxiways, aprons, hardstands, and run-up pads.

4. Distributed Loads

For the case of a superimposed load distributed over an area of considerable extent (Fig. 9-15), the formula for load on the sewer pipe is:

$$W_{sd} = C_s pFB_c \tag{9.14}$$

in which W_{sd} is the load on the sewer pipe, in newtons per unit length (pounds per unit length); p is the intensity of distributed load, in pascals (pounds per square foot); F is the impact factor; B_c is the width of the sewer pipe, in meters (feet); C_s is the load coefficient, a function of $D/(2H)$ and $(M/2H)$ from Table 9-4; H is the height from the top of the sewer pipe to the ground surface, in meters (feet); and D and M are the width and length, respectively, of the area over which the distributed load acts, in meters (feet).

For the case of a uniform load offset from the center of the sewer pipe, the loads per unit length of the sewer pipe may be determined by a combination of rectangles. For determination of the stress below a point such as A in Fig. 9-16, as a result of the loading in the rectangle BCDE, the area may be considered to consist of four rectangles: AJDF−AJCG−AHEF+AHBG. Each of these four rectangles has a corner at point A. By computing $D/2H$ and $M/2H$ for each rectangle, the load coefficient for each rectangle can be taken from Table 9-4. Since point A is at the corner of each rectangle, the load coefficients from Table 9-4 should be divided by 4. A combination of the stresses from the four rectangles, with signs as indicated above, gives the desired stress.

Values of C_s can be read directly from Table 9-4 if the area of the

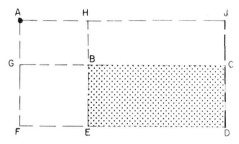

Fig. 9-16. Diagram for obtaining stress at point A caused by load in shaded area BCDE.

distributed superimposed load is centered vertically over the center of the sewer pipe under consideration.

The load on the sewer pipe can be computed by adding algebraically the effect of various rectangles of loaded area if the area of the distributed superimposed load is not centered over the sewer pipe, but is displaced laterally and longitudinally. It is more convenient to work in terms of load under one corner of a rectangular loaded area rather than at the center. Dividing the tabular values of C_s by 4 will give the effect for this condition. Stresses from various types of surcharge loadings can be computed using computer solutions (24,25).

5. Sewer Pipe Under Railway Tracks

The live load may be considered as a uniformly distributed load equal to the weight of locomotive driver axles divided by an area equal to the length occupied by the drivers multiplied by the length of ties. In addition, 2,900 N/m (200 lb/ft) should be allowed for the weight of the track structure.

6. Sewer Pipe Under Rigid Pavement

A method of computing the load transmitted to sewer pipe under rigid pavement is given in the publication, *Vertical Pressure on Culverts Under Wheel Loads on Concrete Pavement Slabs* (10).

Sample Calculations

Example 9-9. Determine the load on a 24-in. diameter pipe under 3 ft of cover caused by a 10,000-lb truck wheel applied directly above the center of the pipe.

Assume the pipe section is 2.5 ft long; the wall thickness is 2 in.; and the impact factor is 1.0. Then $B_c = 24 + 4 = 28$ in. $= 2.33$ ft; $L = 2.5$ ft; and $H = 3.0$ ft. Finally, $B_c/2H = 2.33/6 = 0.39$; and $L/2H = 2.5/6 = 0.41$; the load coefficient is 0.240 from Table 9-4. Substituting in Eq. 9.13:

$$W_{sc} = 0.240 \times \frac{10,000 \times 1.0}{2.5} = 960 \text{ lb/ft } (14,000 \text{ N/m}).$$

Example 9-10. Determine the load on a 48-in. diameter concrete pipe under 6 ft of cover (bottom of ties to top of pipe) resulting from the Cooper E-80 railroad loading.

Assume the pipe wall thickness is 4 in., the locomotive load consists of four 80,000-lb axles spaced 5 ft center-to-center, the impact factor is 1.40, and

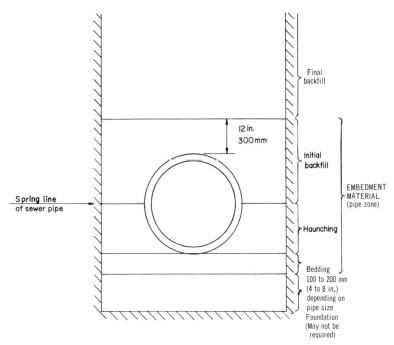

Fig. 9-17. Trench cross section illustrating terminology. Copyright, American Society for Testing and Materials, 1916 Race St., Philadelphia, PA 19103, reprinted with permission.

the weight of track structure is 200 lb/ft. Then $B_c = 48 + 8 = 56$ in. or 4.67 ft; $H = 6$ ft; $D = 8$ ft; and $M = 20$ ft. The unit load plus impact at the base of the ties is:

$$\frac{4 \times 80,000 \times 1.4}{8 \times 20} + \frac{200}{8} = 2,825 \text{ psf}; \quad \frac{D}{2H} = \frac{8}{12} = 0.67;$$

$$\frac{M}{2H} = \frac{20}{12} = 1.67.$$ The influence coefficient is 0.641 (Table 9-4).

From Eq. 9.14:

$$W_{sd} = 0.641 \times 2,825 \times 4.67 = 8,460 \text{ lb/ft (123,420 N/m)}$$

D. PIPE BEDDING AND BACKFILLING

1. General Concepts

The ability of a sewer pipe to safely resist the calculated soil load depends not only on its inherent strength but also on the distribution of the bedding reaction and on the lateral pressure acting against the sides of the sewer pipe.

Construction of the sewer pipe-soil system focuses attention on the pipe zone which is made up of five specific areas: Foundation, bedding, haunching, initial backfill, and final backfill (see Fig. 9-17 for definitions and limits of these five areas). However, all of these areas are not necessarily referred to in all pipe design standards. The discussion in this section is in general terms and is intended to describe the effect of the various areas on the pipe-soil system. More detailed requirements are given in the following sections on pipe design and in Chapter 10.

For all sewer pipe materials, the calculated vertical load is assumed to be uniformly distributed over the width of the pipe. This assumption was originated by Marston's work; the assumption is part of the bedding factors developed by Spangler and presented in this chapter. Many years of field experience indicate the assumption results in conservative designs.

The load-carrying capacity of sewer pipes of all materials is influenced by the sewer pipe-soil system, although the importance of the specific areas may vary with different pipe materials. The information in this chapter includes descriptions of pipe beddings for the following:
- Rigid sewer pipe in trench,
- rigid sewer pipe in embankment,
- flexible non-metallic pipe,
- ductile iron pipe.

Detailed information on pipe bedding classes is contained in the various ANSI and ASTM Specifications or industry literature for each material. The design engineer should consult the applicable specification or literature for information to be used in design.

2. Foundation

The foundation provides the base for the sewer pipe-soil system. In trench conditions, the total weight of the pipe and soil backfill will normally be no more than the weight of the excavated soil. In this case, foundation pressures are not increased from the initial condition, and the designer should be concerned primarily with the presence of unsuitable soils, such as peat or other highly organic or compressible soils, and with maintaining a stable trench bottom.

If the full benefit of the bedding is to be achieved, the bottom of the trench or embankment must be stable. Methods for achieving this condition are discussed in Chapter 10.

3. Bedding

The contact between a pipe and the foundation on which it rests is the sewer pipe bedding. The bedding has an important influence on the distribution of the reaction against the bottom of the sewer pipe, and therefore influences the supporting strength of the pipe as installed.

Some research (11,28) has indicated that a well-graded crushed stone is a more suitable material for sewer pipe bedding than well-graded gravel. Both materials, however, are better suited than a uniformly graded pea gravel.

Even though larger particle sizes give greater stability, the maximum size and shape of granular embedment should also be related to the pipe material and the recommendations of the manufacturer. For example, sharp angular embedment material over 12 to 20 mm (0.5 to 0.75 in.) should not be used against corrosion protection coatings. For small sewer pipes, the maximum size should be limited to about 10% of the pipe diameter.

Soil classifications under the Unified Soil Classification System, including manufactured materials, are grouped into five broad categories according to their ability to develop an interacting sewer pipe-soil system (Table 9-6). These soil classes are described in ASTM D 2321.

Table 9-6 Soil Classifications*

Soil Class (1)	Group Symbol (2)	Typical Names (3)	Comments (4)
I		Crushed rock	angular, 6-40 mm
II	GW GP SW SP	Well graded gravels Poorly graded gravels Well graded sands Poorly graded sands	40 mm maximum
III	GM GC SM SC	Silty gravels Clayey gravels Silty sands Clayey sands	
IV	MH, ML CH, CL	Inorganic silts Inorganic clays	Not recommended for bedding, haunching or initial backfill
V	OL, OH PT	Organic silts and clays Peat	

*For a more detailed description, see ASTM D2321 — "Recommended Practice for Underground Installation of Flexible Thermoplastic Sewer Pipe".
Copyright, American Society for Testing and Materials, 1916 Race St., Philadelphia, PA 19103, reprinted with permission.
Note: 1mm = 0.039 in.

In general, crushed stone or gravel meeting the requirements of ASTM Designation C33, Gradation 67 [19 to 4.8 mm (0.75 in. to No. 4)] will provide the most satisfactory sewer pipe bedding.

In some locations the natural soils at the level of the bottom of the sewer pipe may be sands of suitable grain size and density to serve as both foundation and bedding for the pipe. In such situations, as determined by the design engineer, it may not be necessary to remove and replace these soils with the special bedding materials described above. If the natural soil is left in place, it should be properly shaped for the class of bedding required.

4. Haunching

The soil placed at the sides of a pipe from the bedding up to the spring line is the haunching. The care with which this material is placed has a significant influence on the performance of the sewer pipe, particularly in the space just above the bedding. Poorly compacted material in this space will result in a concentration of reaction at the bottom of the pipe.

For flexible pipe, compaction of the haunching material is essential. For rigid pipe, compaction can ensure better distribution of the forces on the pipe. Material used for sewer pipe haunching should be shovel sliced or otherwise placed to provide uniform support for the pipe barrel and to fill completely all voids under the pipe. Because of space limitations, haunching material is often compacted manually. Results should be checked to verify that the class of bedding or installation criteria are achieved.

Material used for haunching may be crushed stone or sand, or a well-graded granular material of intermediate size. If crushed stone is used, it should be subject to the same size limitation and caution regarding use against corrosion protection coatings. Sand should not be used if the pipe zone area is subject to a fluctuating groundwater table or where there is a possibility of the sand migrating into the pipe bedding or trench walls.

5. Initial Backfill

Initial backfill is the material which covers the sewer pipe and extends from the haunching to some specific point [generally 6 to 12 in. (15 to 30 cm)] above the top of the pipe, depending on the class of bedding. Its function is to anchor the sewer pipe, protect the pipe from damage by subsequent backfill, and ensure the uniform distribution of load over the top of the pipe.

The initial backfill is usually not mechanically tamped or compacted, since such work may damage the sewer pipe, particularly if done over the crown of the pipe. Therefore it should be a material which will develop a uniform and relatively high density with little compactive effort. Initial backfill should consist of suitable granular material, but not necessarily as select a material as that used for bedding and haunching. Clayey materials requiring mechanical compaction should not be used for initial backfill.

The fact that little compaction effort is used on the initial backfill should not lead to carelessness in choice or placement of material. In particular for large sewer pipes, care should be taken in placing both the initial backfill and final backfill over the crown to avoid damage to the sewer pipe.

6. Final Backfill

The choice of material and placement methods for final backfill are related to the site of the sewer line; they generally are not related to the design of the sewer pipe. Under special embankment conditions or induced trench conditions, final backfill may play an important part in the sewer pipe design. However, for most trench installations, final backfill does not affect the pipe design.

The final backfill of trenches in traffic areas, such as under improved existing surfaces, is usually composed of material that is easily densified to minimize future settlement. In undeveloped areas, the final backfill often will consist of the excavated material placed with little compaction and left mounded over the trench to allow for future settlement. Studies have indicated that with some soils, this settlement may continue for over 10 yr.

Trench backfilling should be done in such a way as to prevent dropping of material directly on top of a sewer pipe through any great vertical distance. When placing material with a bucket, the bucket should be lowered so that the shock of falling earth will not cause damage.

E. DESIGN SAFETY FACTOR AND PERFORMANCE LIMITS

1. General Concepts

In the design of sanitary sewer pipes, the selection of a factor of safety is an essential element of the structural design requirements. When the structural design requirements are defined for a given construction material, the most severe or maximum performance limits of that material as related to the proposed service or design application limit must be determined; the ratio of the two limits is the "factor of safety." Maximum performance limits and design values may be defined in terms of strength, stress, strain, or product deformation depending on the characteristics of the material under consideration. The selected factor of safety is applied to the maximum performance limit to calculate a lower value which is then used as a design or service performance value.

The selection of a desired factor of safety is essentially based on a risk/value assessment which must relate to the specific conditions anticipated in an application, the failure mode of the construction material, and the potential cost of system failure. Factors of safety compensate for unexpected construction deficiencies. They should not be relied on to compensate for poor construction practice or for inadequate inspection. Properly established design performance values, including adequate factors of safety, must be realized in installation and operation to provide reasonable assurance of acceptable long-term system performance.

The relationship between safety factors and design performance values as related to structural requirements for a sanitary sewer pipe, is similar for rigid and flexible sewer pipe. However, an understanding of the differences between design requirements for rigid sewer pipe and the requirements for flexible sewer pipe system design is important.

Recommended field procedures and safety factors are discussed below.

2. Rigid Sewer Pipe

Design performance limits for rigid sewer pipes generally are expressed in terms of strength under load. There are two alternative methods of determining the service strength of reinforced concrete sewer pipe — by "strength design" analysis, or by testing. The design method more accurately represents the relative performance of pipe in the field under the service load. However, testing is generally used for all types of precast or prefabricated rigid pipe.

Strengths of rigid sewer pipe usually are measured in terms of the ultimate three-edge bearing strength, and of ultimate and 0.3-mm (0.01-in.) crack, three-edge bearing strengths for reinforced concrete sewer pipe. A factor of safety of 1.0 should be applied to the 0.3 mm (0.01-in.) crack load for reinforced concrete sewer pipe, and of 1.25 to 1.50 should be applied to the specified minimum ultimate three-edge bearing strength to determine the working strength for other rigid pipes. Common practice is to use a factor of safety of 1.25 for the ultimate load of reinforced concrete sewer pipe, and up to 1.50 for vitrified clay.

3. Flexible Sewer Pipe

Design performance limits for flexible sewer pipes may be expressed in terms of stress or strain in the pipe wall, crushing or buckling in the pipe wall, or deflection. The most common limitation is deflection. The deflection limitation is established as a design performance limit to provide a factor of safety, against structural failure or any type of distress which might tend to limit the service life of the pipe. This design limit will vary with different pipe materials and the pipe manufacturing process. Deflection limits of up to 7.5% have been used for the design of certain types of flexible pipe. Pipes must be able to deflect without cracking, liner failure, or other distress, and they should be designed accordingly with a reasonable factor of safety.

Certain types of plastic pipes are subject to strain deterioration. There is a limiting strain which, if exceeded over a period of time, will eventually result in cracking of the pipe wall, particularly if exposed to an acid environment.

Manufacturers should be consulted on the value of this limiting strain.

4. Recommendations for Field Procedures

The factor of safety against ultimate collapse of sewer pipe is about the same as that used in the design of most engineered structures of monolithic concrete. However, the design of sewer pipe is based on calculated loads, bedding factors, and experimental factors, which are less well-defined than the dead and live loads used in building design. It is, therefore, important that the loads imposed on the sewer pipe be not greater than the design loads.

To obtain the objective of imposed loads being less than design loads, the following procedures are recommended:

(1) *Specifications*. Construction specifications should set forth limits for the width of trench below the top of sewer pipe. The width limits should take into account the minimum width required to lay and join sewer pipe and the maximum allowable width for each class of sewer pipe and bedding to be used. Where the depth is such that a positive projecting condition will be obtained, maximum width should be specified as unlimited unless the width must be controlled for some reason other than to meet structural requirements of the sewer pipe. Appropriate corrective measures should be specified in the event the maximum allowable width is exceeded. These measures may include provision for a higher class of bedding or concrete encasement. Maximum allowable construction live loads should be specified for various depths of cover if appropriate for the project.

(2) *Inspection*. Construction should be observed by an experienced engineer or inspector who reports to a competent field engineer.

(3) *Testing*. Sewer pipe testing should be under the supervision of a reliable testing laboratory, and close liaison should be maintained between the laboratory and the field engineer.

(4) *Field Conditions*. The field engineer should be furnished with sufficient design data to enable him to evaluate unforeseen conditions intelligently. The field engineer should be instructed to confer with the design engineer if changes in design appear advisable.

(5) *Sheeting*. Where sheeting is to be removed, pulling should be done in stages. The space formerly occupied by the sheeting must be backfilled completely, and field inspection should verify this.

a. Effect of Trench Sheeting

Because of the various alternative methods employed in sheeting trenches, generalizations on the proper construction procedure to follow to ensure that the design load is not exceeded are risky and dangerous. Each method of sheeting and bracing should be studied separately. The effect of a particular system on the sewer pipe load, as well as the consequences of removing the sheeting or the bracing, must be estimated.

It is difficult to obtain satisfactory filling and compaction of the void left when wood sheeting is pulled. Wood sheeting driven alongside the sewer pipe should be cut off and left in place to an elevation of 450 mm (1.5 ft) above the top of the sewer pipe.

If granular materials are used for backfill it is possible to fill and compact the voids left by the wood sheeting if the material is placed in lifts and jetted as the sheeting is pulled. If cohesive materials are used for backfill, a void will be left on pulling the wood sheets and the full weight of the prism of earth contained between the sheeting will come to bear on the sewer pipe.

Skeleton sheeting or bracing should be cut off and left in place to an elevation of 450 mm (1.5 ft) over the top of the sewer pipe if removal of the trench support might cause a collapse of the trench wall and a widening of the trench at the top of the conduit. Entire skeleton sheeting systems should be left in place if removal would cause collapse of the trench before backfill can be placed.

Where steel soldier beams with horizontal lagging between the beam flanges are used for sheeting trenches, efforts to reclaim the steel beams before the trench is backfilled may damage pipe joints. It is recommended that use of this type of sheeting be allowed only on stipulation that the beams be pulled after backfilling and the lagging be left in place.

Steel sheeting may be used and reused many times, and the relative economy of this type of sheeting compared with timber or timber and soldier beams should be explored. Because of the thinness of the sheeting it is often feasible to achieve reasonable compaction of backfill so that the steel sheeting may be withdrawn with about the same factor of safety against settlement of the surfaces adjacent to the trench as that for other types of sheeting left in place.

b. Trench Boxes

Several backfilling techniques are possible when a trench box is used. Granular material can be placed between the box and the trench wall immediately after placing the box, and the box is advanced by lifting slightly before moving forward. Boxes can be made with a step at the rear which makes the trench wall accessible for compacting the embedment against the walls. Also, the box can be used where the sewer pipe is laid in a small sub-trench, and the box is utilized only in the trench above the top of the sewer pipe. Advancement of the box must be done carefully to avoid pulling pipe joints apart.

c. Pipe Bedding and Embedment

To assure that the sewer pipe is properly bedded or embedded it is suggested that compaction tests be made at selected or critical locations, or that the method of material placement be observed and correlated to known results.

Where compaction measurement or control is desired or required, the recommended references are:
- ASTM D2049, "Standard Method of Test for Relative Density of Cohesionless Soils"
- ASTM D698, "Standard Method of Test for Moisture Density Relations of Soils Using 5.5-lb (2.5 kg) Rammer and 12-in (204.8 mm) Drop"
- ASTM D2167, "Standard Method of Test for Density of Soil in Place by the Rubber-Balloon Method"
- ASTM D1556, "Standard Method of Test for Density of Soil in Place by the Sand-Cone Method"
- ASTM D2922, "Standard Method of Test of Density of Soil and Soil-Aggregate in Place by Nuclear Methods (Shallow Depth)."

It is recommended that the in-place density of Class I and Class II embedment materials be measured by ASTM D2049 by percentage of relative density, and Class III and Class IV measured by either ASTM D2167, D1556 or D2922, by percentage of Standard Proctor Density according to ASTM D698 or AASHTO T99.

An economical indicator of proper bedding and backfill execution on flexible sewer pipes may be by testing the inside pipe deflection by a go/no-go mandrell. Mandrell dimensions can be determined by applying the initial

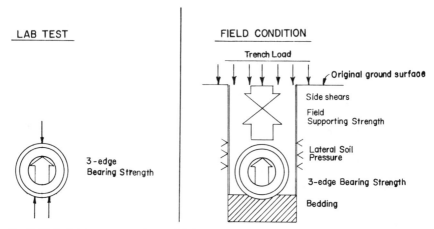

Fig. 9-18. Both laboratory testing and field conditions should be used for rigid sewer pipe strength determination.

deflection to the base inside diameter as determined from the appropriate ASTM pipe standard. The initial deflection does not include the deflection lag factor, as is applied to ultimate long-term design deflection.

F. RIGID SEWER PIPE DESIGN

1. General Relationships

The inherent strength of a rigid sewer pipe usually is given by its strength in the three-edge bearing test. Although this test is both convenient and severe, it does not reproduce the actual field load conditions. Thus to select the most economical combination of bedding and sewer pipe strength, a relationship must be established between calculated load, laboratory strength, and field strength for various installation conditions (Fig. 9-18).

Field strength, moreover, depends on the distribution of the reaction against the bottom of the sewer pipe and on the magnitude and distribution of the lateral pressure acting on the sides of the pipe. These factors, therefore, make it necessary to qualify the term "field strength" with a description of conditions of installation in a particular case, as they affect the distribution of the reaction and the magnitude and distribution of lateral pressure.

Just as for sewer pipe load computations, it is convenient when determining field strength to classify installation conditions as either trench or embankment.

2. Laboratory Strength

Rigid sewer pipe is tested for strength in the laboratory by the three-edge bearing test (Fig. 9-18). Methods of testing are described in detail in ASTM Specification C 301 for vitrified clay sewer pipe, C 497 for concrete and reinforced concrete sewer pipe, and C 500 for asbestos cement sewer pipe.

The minimum strengths required for the three-edge bearing tests of the various types of rigid sewer pipe are stated in the ASTM Specifications for the pipe.

In the case of reinforced concrete sewer pipe, laboratory strengths are divided into two categories: The load that will produce a 0.3-mm (0.01-in.)

crack, and the ultimate load the sewer pipe will withstand. According to ASTM C 76, the ultimate strength load in three-edge bearing is 1.50 to 1.25 times greater than the load that will produce a 0.3-mm (0.01-in) crack, and the factor of safety used in design depends on which load is being considered.

Asbestos cement, non-reinforced concrete, and clay sewer pipe class designations indicate the ultimate strength in three-edge bearing directly in newtons per meter (pounds per foot) for any diameter. A Class 1500 asbestos cement sewer pipe, for example, must have a laboratory ultimate strength of 22,000 N/m (1,500 lb/ft) for every diameter in that class.

3. Design Relationships

The structural design of rigid sewer pipe systems relates to the product's performance limit, expressed in terms of strength of the installed sewer pipe. Based on anticipated field loadings and concrete and reinforcing steel properties, calculations of shear, thrust, and moment can be made and compared to the ultimate strength of the section. This method is commonly used for reinforced concrete cast-in-place sewer pipes.

For precast sewer pipes, the design strength is commonly related to a three-edge bearing test strength measured at the manufacturing plant, as described in the following text. The plant test is much more severe because it develops shearing forces in the pipe wall that usually are not encountered in the field. Special reinforcement to resist shear in testing may be required in heavily reinforced sewer pipe, which would not be required by the more uniformly distributed field loading. In either case, the design load is equal to the field strength divided by the factor of safety. The design load is the calculated load on the pipe, and the field strength is the ultimate load which the pipe will support when installed under specified conditions of bedding and backfilling.

The field strength is equal to the three-edge bearing test strength of the pipe times the bedding factor. The ratio of the strength of a sewer pipe under any stated condition of loading and bedding to its strength measured by the three-edge bearing test is called the bedding factor.

Bedding factors for trenches and embankments were determined experimentally between the years 1925 to 1935 at Iowa State College. They can also be calculated for most classes of bedding (23). The experimentally determined and calculated relationships may be combined as follows to determine the required bedding factor for any given pipe class:

$$\text{Bedding Factor} = \frac{\text{Design Load} \times \text{Factor of Safety}}{\text{Three-Edge Bearing Strength}}$$

By rearranging this expression, the required three-edge bearing strength for a given class of bedding can be calculated as follows:

$$\text{Required Three-Edge Bearing Strength} = \frac{\text{Design Load} \times \text{Factor of Safety}}{\text{Bedding Factor}}$$

The strength of reinforced concrete sewer pipe in newtons per meter (pounds per foot) at either the 0.3-mm (0.01-in.) crack or ultimate load, divided by the nominal internal diameter of the sewer pipe in millimeters (feet), is defined as the D-load strength: Different classes of reinforced concrete sewer pipe per ASTM C 76 and other reinforced concrete pipe specifications are defined by D-load strengths. For example, ASTM C 76 Class IV Reinforced Concrete Sewer Pipe is manufactured to a D-load of 100 N/m/mm of diameter (2,000 lb/ft/ft) to produce the 0.3-mm (0.01-in) crack, and 150 N/m/mm (3,000 lb/ft/ft) to produce the ultimate load. Consequently, a 1.2-m (48-in.) diameter Class IV Reinforced Concrete Sewer Pipe per ASTM C 76 would have a minimum laboratory strength of 120,000 N/m (8,000 lb/ft) to produce the 0.3-mm (0.01-in) crack and 180,000 N/m (12,000 lb/ft) at ultimate strength.

$$\text{D-Load} = \frac{\text{Design Load} \times \text{Safety Factor}}{\text{Bedding Factor} \times \text{Diameter}}$$

In considering the design requirements, the design engineer should evaluate the options of different types of rigid sewer pipe and different bedding classes, keeping in mind the differences in the relationship between test loadings and field loadings. It is customary when designing for 0.3-mm (0.01-in.) crack loads to use a factor of safety of 1.0, and when designs are based on ultimate strengths to use a factor of safety varying between 1.25 and 1.50.

4. Rigid Sewer Pipe Installation—Classes of Bedding and Bedding Factors for Trench Conditions

Four classes of beddings are used most often for sewer pipes in trenches (Fig. 9-19). They are described in the following subsections.

a. Class A — Concrete Cradle

The sewer pipe is bedded in a cast-in-place cradle of plain or reinforced concrete having a thickness equal to one-fourth the inside pipe diameter, with a minimum of 100 mm (4 in.) and a maximum of 380 mm (15 in.) under the pipe barrel and extending up the sides for a height equal to one-fourth the outside diameter. The cradle shall have a width at least equal to the outside diameter of the sewer pipe barrel plus 200 mm (8 in.). Construction procedures must be executed carefully to prevent the sewer pipe from floating off line and grade during placement of the cradle concrete.

If the cradle is made of reinforced concrete, the reinforcement is placed transverse to the pipe and 8 mm (3 in.) clear from the bottom of the cradle. The percentage of reinforcement, p, is the ratio of the area of transverse reinforcement to the area of concrete cradle at the pipe invert above the centerline of the reinforcement.

Consideration must be given to the points at which the cradle (or arch, in the next section) begins and terminates with respect to the pipe joints. In general, the concrete cap, cradle, or envelope starts and terminates at the face of a pipe bell or collar to avoid shear cracks.

Haunching and initial backfill above the cradle to 300 mm (12 in.) above the crown of the sewer pipe should be placed and compacted as described in sections D4 and D5 of this chapter. The cradle must be cured sufficiently to develop full bedding support prior to final backfilling.

The bedding factor for Class A concrete cradle bedding is 2.2 for plain concrete with lightly tamped backfill; 2.8 for plain concrete with carefully

STRUCTURAL REQUIREMENTS

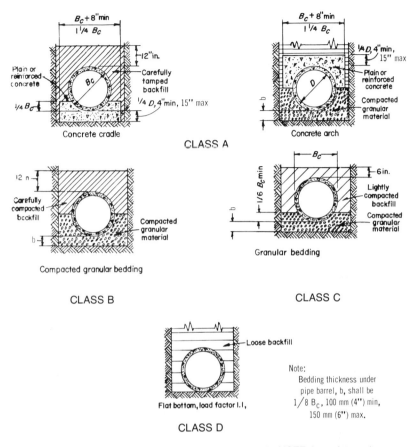

Fig. 9-19. Classes of bedding for rigid sewer pipes in trench. NOTE: In rock trench, excavate at least 15 cm (6 in.) below bell of pipe except where concrete cradle is used (in. × 25.4 = mm).

tamped backfill; up to 3.4 for reinforced concrete with the percentage of reinforcement, p, equal to 0.4%, and up to 4.8 with p equal to 1.0%.

Class A — Concrete Arch

The sewer pipe is bedded in carefully compacted granular material having a minimum thickness of one-eighth the outside sewer pipe diameter but not less than 100 mm (4 in.) or more than 150 mm (6 in.) between the sewer pipe barrel and bottom of the trench excavation. Granular material is then placed to the spring line of the sewer pipe and across the full breadth of the trench. The haunching material beneath the sides of the arch must be compacted so as to be unyielding. Crushed stone in the 5-mm to 20-mm (0.25-in. to 0.75-in.) size range is the preferred material. The top half of the sewer pipe is covered with a cast-in-place plain or reinforced concrete arch having a minimum thickness of 100 mm (4 in.) or one-fourth the inside pipe diameter but not to exceed 380 mm (15 in.), and having a minimum width equal to the outside sewer pipe diameter plus 200 mm (8 in.).

If the arch is made of reinforced concrete, the reinforcement is placed transverse to the pipe and 50 mm (2 in.) clear from the top of the arch. The percentage of reinforcement is the ratio of the transverse reinforcement to the area of concrete arch above the top of the pipe and below the centerline of the reinforcement.

The bedding factor for Class A concrete arch-type bedding is 2.8 for plain concrete; up to 3.4 for reinforced concrete with p equal to 0.4%; and up to 4.8 for p equal to 1.0%.

b. Class B Bedding

The sewer pipe is bedded in carefully compacted granular material. The granular bedding has a minimum thickness of one-eighth the outside sewer pipe diameter but not less than 100 mm (4 in.) or more than 150 mm (6 in.), between the barrel and the trench bottom, and covering the full width of the trench.

The haunch area of the sewer pipe must be fully supported; therefore, the granular material should be shovel sliced or otherwise compacted under the pipe haunch. Both haunching and initial backfill to a minimum depth of 300 mm (12 in.) over the top of the sewer pipe should be placed and compacted.

The bedding factor for Class B bedding is 1.9.

c. Class C Bedding

The sewer pipe is bedded in compacted granular material. The bedding has a minimum thickness of one-eighth the outside sewer pipe diameter, but not less than 100 mm (4 in.) or more than 150 mm (6 in.), and shall extend up the sides of the sewer pipe one-sixth of the pipe outside diameter. The remainder of the sidefills, to a minimum depth of 150 mm (6 in.) over the top of the pipe, consists of lightly compacted backfill.

The bedding factor for Class C bedding is 1.5.

d. Class D Bedding

For this class of bedding the bottom of the trench is left flat, as cut by excavating equipment. Successful Class D installations of rigid conduits can be achieved in locations with appropriate soil conditions, trench width control, and light superimposed loads. Care must be taken to prevent point loading of sewer pipe bells; the excavation of bell holes will prevent such loading. Existing soil should be shovel sliced or otherwise compacted under the haunching of the sewer pipe to provide some uniform support. In poor soils, granular bedding material is generally a more practical, cost effective installation.

Since the field conditions with Class D bedding can approach the load conditions for the three-edge bearing test, the bedding factor for Class D bedding is 1.1.

5. Encased Pipe

Total encasement of rigid sewer pipe in concrete may be necessary where the required field strength cannot be obtained by other bedding methods.

A typical concrete encasement detail is shown in Fig. 9-20. The bedding factor for concrete encasement varies with the thickness of concrete and the use of reinforcing and may be greater than that for a concrete cradle or arch. Concrete thickness and reinforcing should be determined by the application of conventional structural theory and analysis. The bedding factor for the encase-

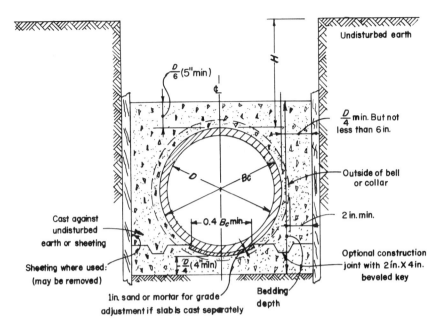

Fig. 9-20. Typical concrete encasement details (in. × 25.4 = mm).

ment shown in Fig. 9-20 is 4.5.

Concrete encasement also may be required for sanitary sewers built in deep trenches to ensure uniform support or for sewer pipe built on comparatively steep grades where there is the possibility that earth beddings may be eroded by currents of water under and around the pipe. Flotation of the pipe during concrete placement should be prevented by suitable means.

Sample Calculations:

Example 9-11. Refer to *Example 9-1*, which describes the backfill load on a 24-in. diameter pipe with 14 ft of cover; the load was found to be 4,720 lb/ft. If vitrified clay pipe is to be specified and a factor of safety of 1.5 is selected, the design load will be 4,720 × 1.5 or 7,080 lb/ft (103,300 N/m).

The crushing strength requirement of 24-in. diameter extra-strength clay sewer pipe (ASTM C 700) based on the three-edge bearing method is 4,400 lb/ft. Dividing this into 7,080 lb/ft (the design load), the minimum required bedding factor of 1.61 is obtained. Fig. 9-19 indicates that a Class B bedding is required for this installation.

Example 9-12. Refer to *Example 9-3*. If a 24-in. diameter reinforced concrete sewer pipe (one line of reinforcement near center of wall) and a factor of safety of 1.50 based on the minimum ultimate test strength are selected, the design load will be 1.50 × 5,750 or 8,620 lb/ft (125,900 N/m).

If the minimum ultimate test strength of the pipe is 6,000 lb/ft (ASTM C 76, Class IV-3,000 D), the required bedding factor will be 8,620/6,000 or 1.44. According to Fig. 9-19, this installation will require a Class C bedding.

Example 9-13. Assume the 48-in. diameter pipe in *Example 9-5* is bedded on an unreinforced concrete cradle, with p equal to m equal to 0.5, and k equal to 0.33. Using a factor of safety of 1.5 based on the ultimate three-edge bearing strength of the pipe:

$\dfrac{H}{B_c} = 6.63; C_c = 10.0; x = 0.856;$ and $N = 0.575$

From Eq. 9.16:

$$q = \dfrac{0.5 \times 0.33}{10.0}(6.63 + 0.25) = 0.114$$

and from Eq. 9.15:

$$L_f = \dfrac{1.431}{0.575 - (0.856 \times 0.114)} = 3.01$$

The required three-edge bearing strength at ultimate load is:

$$\dfrac{29{,}100 \times 1.5}{3.01} = 14{,}500 \text{ lb/ft (211,700 N/m), or 3,630 D (175 D in N/m/mm diam)}.$$

Therefore ASTM C 76, Class V pipe should be used.

6. Field Strength in Embankments

It is possible for the active soil pressure against the sides of rigid sewer pipe placed in an embankment to be a significant factor in the resistance of the structure to vertical load. This factor is important enough to justify a separate examination of the field strength of embankment sewer pipe.

The following discussion of field strength in embankments is based on theory developed by Marston and Spangler. The design engineer, however, should approach designs based on this theory with some caution. With time, lateral pressures for trench installation usually will approach "at rest" conditions, which correspond to vertical overburden times a factor (usually between 0.5 and 1.0). However, for a negative projecting conduit, a positive projecting conduit, and induced trench embankment conditions, the lateral pressure magnitude and distribution may be much different, and these may control the structural design of the sewer pipe.

a. Positive Projecting Sewer Pipe

The bedding factor for rigid sewer pipes installed as projecting sewer pipe under embankments or in wide trenches depends on the class of bedding in which the sewer pipe is laid, the magnitude of the active lateral soil pressure against the sides of the sewer pipe, and the area of the sewer pipe over which the active lateral pressure is effective.

For projecting sewer pipe, the bedding factor, L_f, is:

$$L_f = \dfrac{A}{N - xq} \qquad (9.15)$$

STRUCTURAL REQUIREMENTS

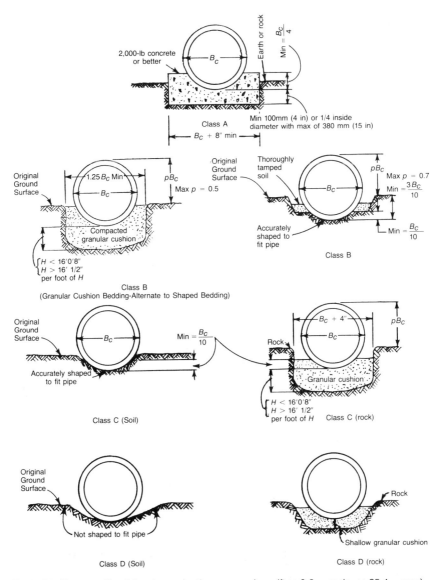

Fig. 9-21. Classes of bedding for projecting sewer pipes (ft × 0.3 = m; in. × 25.4 = mm).

Table 9-7 Values of A for Circular, Elliptical and Arch Sewer Pipe

Sewer Pipe Shape (1)	A (2)
Circular	1.431
Elliptical:	
Horizontal elliptical and arch	1.337
Vertical elliptical	1.021

Table 9-8 Values of N

Class of Bedding (1)	Value of N		
	Sewer Pipe Shape		
	Circular (2)	Horizontal Elliptical (3)	Vertical Elliptical (4)
A (reinforced cradle)	0.421 to 0.505	—	—
A (unreinforced cradle)	0.505 to 0.636	—	—
B	0.707	0.630	0.516
C	0.840	0.763	0.615
D	1.310	—	—

in which A is a sewer pipe shape factor; N is a parameter which is a function of the bedding class; x is a parameter dependent on the area over which lateral pressure effectively acts; and q is the ratio of total lateral pressure to total vertical load on the sewer pipe.

Classes of bedding for projecting sewer pipe are shown in Fig. 9-21. The values of A for circular, elliptical, and arch sewer pipe are shown in Table 9-7.

Values of N for various classes of bedding are given in Table 9-8. Values of x for circular, elliptical and arch sewer pipe are listed in Table 9-9. The projection ratio, m, refers to the fraction of the sewer pipe diameter over which lateral pressure is effective. For example, if lateral pressure acts on the top half of the sewer pipe above the horizontal diameter, m equals 0.5. The ratio of total lateral pressure to total vertical load, q, for positive projecting sewer pipe may be estimated by the formula:

$$q = \frac{mk}{C_c}\left(\frac{H}{B_c} + \frac{m}{2}\right) \qquad (9.16)$$

Table 9-9 Values of x

Fraction of Sewer Pipe Subjected to Lateral Pressure, m (1)	x			
	Class A Bedding	Other then Class A Bedding		
	Circular Sewer Pipe (2)	Circular Sewer Pipe (3)	Horizontal Elliptical Sewer Pipe (4)	Vertical Elliptical Sewer Pipe (5)
0	0.150	0	0	0
0.3	0.743	0.217	0.146	0.238
0.5	0.856	0.423	0.268	0.457
0.7	0.811	0.594	0.369	0.639
0.9	0.678	0.655	0.421	0.718
1.0	0.638	0.638	—	—

in which k is the ratio of unit lateral pressure to unit vertical pressure (Rankine's ratio). A value of k equal to 0.33 usually will be sufficiently accurate for use in Eq. 9.16. Values of C_c are found in Fig. 9-7.

b. Negative Projecting Sewer Pipe

The bedding factor for negative projecting sewer pipe may be the same as that for trench conditions corresponding to the various classes of bedding given in section F4 of this chapter. The bedding factors for Class B, C, and D bedding do not take into account lateral pressures against the sides of the sewer pipe, because unfavorable construction conditions often prevail at the bottom of a trench. However, in the case of negative projecting sewer pipe, conditions may be more favorable and it may be possible to compact the sidefill soils to the extent that some lateral pressure against the sewer pipe can be relied on. If such favorable conditions are anticipated, it is suggested that the bedding factor be computed by means of Eqs. 9.15 and 9.16 using a value of k equal to 0.15 for estimating the lateral pressure on the sewer pipe.

c. Induced Trench Conditions

Induced trench sewer pipes usually are installed as positive projecting sewer pipes before the overlying soil is compacted and the induced trench is excavated. Therefore, lateral pressures are effective against the sides of the sewer pipe, and the bedding factor should be calculated using Eqs. 9.15 and 9.16.

G. FLEXIBLE SEWER PIPE DESIGN

Several types of flexible pipe are available for use as sanitary sewer pipe material. Among the most common are ductile iron pipe (DIP), acrylonitrile-butadiene-styrene (ABS) composite pipe and (ABS) solid wall pipe, polyvinyl chloride (PVC) pipe, polyethylene (PE) pipe, fiberglass-reinforced plastics (FRP) and reinforced plastic mortar (RPM) pipe. Coated corrugated metal pipe (CMP) has been used for sanitary sewer pipe, but its use is not common.

1. General Method

Flexible sewer pipes under earth fills and in trenches derive their ability to support load from their inherent strength plus the passive resistance of the soil as the pipe deflects and the sides of the sewer pipe move outward against the soil sidefills. Proper compaction of the soil sidefills is important to the long term structural performance of flexible sewer pipe. This type of pipe fails by excessive deflection and by collapse, buckling, cracking or delamination rather than by rupture of the sewer pipe walls as in the case of rigid pipes. The extent to which flexible pipe deflects is most commonly used to judge performance and as a basis for design. The amount of deflection considered permissible is dependent on physical properties of the pipe material used and project limitations.

The limiting buckling stress for flexible pipes takes into account the restraining effect of the soil structure around the pipe and the properties of the pipe wall. Equations for the critical stress in the pipe wall can be found in manufacturers' handbooks for the various types of pipe.

Empirical data on long-term deflection in different burial conditions generally can be obtained from the manufacturer. When such design data are not available, the approximate long-term deflection of flexible sewer pipe can be calculated using the Modified Iowa Formula developed by Spangler and

Watkins:

$$\Delta X = \frac{D_L K_b W_c r^3}{EI + 0.061 E'r^3} \tag{9.17}$$

where ΔX = horizontal deflection, in millimeters (inches); K_b = bedding factor; D_L = deflection lag factor; W_c = load, in newtons per linear millimeter (pounds per linear inch); r = mean radius of pipe, in millimeters (inches); E = modulus of tensile elasticity, in pascals (pounds per square inch); I = moment of inertia per length, in millimeters (inches) to fourth power per millimeter (inch); and E' = modulus of soil reaction, in pascals (pounds per square inch).

For small deflections, the vertical deflection ΔY may be assumed to approximately equal the horizontal deflection ΔX in Eq. 9.17. Much research is being conducted on the application of Eq. 9.17 to pipe having a low ratio of pipe stiffness to soil stiffness where the vertical deflection may not be assumed to approximately equal the horizontal deflection. The result of this research is beyond the scope of this chapter but should be reviewed by the designer as a part of the design process.

Eq. 9.17 was developed primarily for flexible corrugated metal sewer pipe under embankments but has been shown to be applicable to most flexible pipe materials.

Flexible sewer pipes that are to support a fill should not be placed directly on a cradle or pile bents. If such supports are necessary, they should have a flat top and be covered with a compressible earth cushion. In those instances where flexible pipe is to be encased in concrete, the pipe manufacturer should be consulted. Flexible pipes should not be encased in concrete unless the encasement is designed for the full vertical load as a rigid pipe.

Recommended values for E' in Eq. 9.17 have been developed and are given in Table 9-10 (22).

The deflection lag factor, empirically determined, compensates for the time consolidation characteristics of the soil, which may permit flexible sewer pipes to continue to deform for some period after installation. Long-term deflection will be greater with light or moderate degrees of compaction of sidefills when compared to values for heavy compaction. The better the compaction, the lower the initial deflection, and the greater the magnitude of the long-term lag factor. Lag factors over 2.5 have been recorded in dry soil. Recommended values of this factor range from 1.25 to 1.50. Bedding requirements for flexible sewer pipe installation are discussed below.

Values of the bedding constant K_b, depending on the width of the sewer pipe bedding, are shown in Table 9-11.

The passive resistance of the soil at the sides of the pipe greatly influences flexible pipe deflection. This passive resistance is expressed as the Modulus of Soil Reaction, E'. It is generally recognized as being related to the degree of compaction of the soil and to the type of soil. Extensive laboratory and field tests by the U.S. Bureau of Reclamation have resulted in the establishment of an empirical relationship between Modulus of Soil Reaction values, degree of compaction of bedding, and type of bedding material (Table 9-10). (See Table 9-6 for soil symbol definition.)

In the deflection formula (Eq. 9.17), the first term in the denominator, EI (stiffness factor), reflects the influence of the inherent stiffness of the sewer pipe on deflection. The second term, $0.061 E'r^3$, reflects the influence of the passive pressure on the sides of the pipe. With flexible pipes, the second term is normally predominant.

Table 9-10 Bureau of Reclamation Average Values of E' for Iowa Formula (for Initial Flexible Pipe Deflection)

Soil type-pipe bedding material (Unified Classification System[a]) (1)	Dumped (2)	E' for Degree of Compaction of Bedding, in pounds per square inch		
		Slight, <85% Proctor, <40% relative density (3)	Moderate, 85%-95% Proctor, 40%-70% relative density (4)	High >95% Proctor, >70% relative density (5)
Fine-grained Soils (LL > 50)[b] Soils with medium to high plasticity CH, MH, CH-MH	No data available; consult a competent soils engineer: Otherwise use $E' = 0$			
Fine-grained Soils (LL < 50) Soils with medium to no plasticity CL, ML, ML-CL, with less than 25% coarse-grained particles	50	200	400	1,000
Fine-grained Soils (LL < 50) Soils with medium to no plasticity CL, ML, ML-CL, with more than 25% coarse-grained particles Coarse-grained Soils with Fines GM, GC, SM, SC[c] contains more than 12% fines	100	400	1,000	2,000
Coarse-grained Soils with Little or No Fines GW, GP, SW, SP[c] contains less than 12% fines	200	1,000	2,000	3,000
Crushed Rock	1,000	3,000	3,000	3,000
Accuracy in Terms of Percentage Deflection[d]	±2	±2	±1	±0.5

[a]ASTM Designation D-2487, USBR Designation E-3.
[b]LL = Liquid limit.
[c]Or any borderline soil beginning with one of these symbols (i.e., GM-GC, GC-SC).
[d]For ± 1% accuracy and predicted deflection of 3%, actual deflection would be between 2% and 4%.

Note: Values applicable only for fills less than 50 ft (15 m). Table does not include any safety factor. For use in predicting initial deflections only, appropriate Deflection Lag Factor must be applied for long-term deflections. If bedding falls on the borderline between two compaction categories, select lower E' value or average the two values. Percentage Proctor based on laboratory maximum dry density from test standards using about 12,500 ft-lb/cu ft (598,000 J/m^3) (ASTM D-698, AASHO T-99, USBR Designation E-11). 1 psi = 6.9 KPa.

Table 9-11 Values of Bedding Constant, K_b

Bedding Angle, in degrees (1)	K_b (2)
0	0.110
30	0.108
45	0.105
60	0.102
90	0.096
120	0.090
180	0.083

Note: 1 deg = 0.017 rad

It should be noted that the E' values in Table 9-10 are average values. Using them results in a 50% chance that the actual deflections will be higher than calculated. A conservative approach would be to use 75% of the E' values given to calculate maximum deflections.

The performance of the flexible sewer pipe in retaining its shape and integrity is largely dependent on the selection, placement, and compaction of the envelope of soil surrounding the structure. For this reason as much care should be taken in the design of the bedding and initial backfill as is used in the design of the sewer pipe. The backfill material selected preferably should be of a granular nature to provide good shear characteristics. Cohesive soils are generally less suitable because of the importance of proper moisture content and the difficulty of obtaining proper compaction in a limited work space.

If, under embankment conditions, the material placed around the sewer pipe is different from that used in the embankment, or if for construction reasons fill is placed around the sewer pipe before the embankment is built, the compacted backfill should cover the pipe by at least 0.3 m (1 ft) and extend at least one diameter to either side.

2. Design of Plastic Sewer Pipes

A wide range of physical properties is available from various plastic sewer pipe materials. Many have ASTM or ANSI specifications.

Thermoplastic pipe materials (ABS, PE, and PVC) are all affected by temperature. Specified test requirements are applicable at 23°C (73.4°F). At higher temperatures the plastic becomes less rigid but impact strength is increased. Lower temperatures result in increased brittleness but greater pipe stiffness. Pipe being installed at low temperatures requires careful handling. Certain chemicals will increase the possibility of brittle fracture for thermoplastics as discussed in Chapter 8.

Thermosetting reinforced plastic pipe is made from a variety of resins, fiberglass reinforcing systems, and aggregates in the structural wall. Manholes and other appurtenances are also made from these materials. Reinforced plastic mortar pipe made in conformance with ASTM D 3262 is also widely used.

Thermosetting resins are temperature sensitive and are affected by ultraviolet radiation. Thermosetting reinforced plastic mortar pipe may have a liner of either thermosetting or thermoplastic resin. It has been found that thermosetting reinforced plastic pipe may be more susceptible to corrosion in certain environments in the deflected position. Tests to determine these properties can be made in accordance with ASTM D 3681 and D 3262, Appendix X2.

The wide range of physical properties of plastics makes possible the production of both flexible and rigid pipe, depending on the materials used. Plastic pipe technology is developing rapidly and new methods for structural design can be expected.

a. Laboratory Load Test

The standard test to determine "pipe stiffness" or load deflection characteristics of plastic pipe is the parallel-plate loading test. This test is conducted in accordance with ASTM D 2412, "Standard Test Method for External Loading Properties of Plastic Pipe by Parallel-Plate Loading."

In the test a short length of pipe is loaded between two rigid parallel flat plates as they are moved together at a controlled rate. Load and deflection are noted.

The parallel-plate loading test determines the pipe stiffness (PS) at a prescribed deflection (ΔY), which for convenience in testing is arbitrarily set at 5%. This is not to be considered a limitation on field deflection. The pipe stiffness is defined as the value obtained by dividing the force (F) per unit length by the resulting deflection in the same units at the prescribed percentage deflection, and is expressed in newtons per millimeter (pounds per inch):

$$PS = \frac{F}{\Delta Y} = \frac{EI}{0.149r^3} \qquad (9.18)$$

where E = modulus of elasticity, in pascals (pounds per square inch); $I = t^3/12$; r = mean radius of pipe, in millimeters (inches); and t = wall thickness, in millimeters (inches).

Minimum required pipe stiffness values are stated in plastic sewer pipe specifications. Table 9-12 lists the ASTM Specifications for the various types of plastic pipe and the corresponding pipe stiffness values.

The stiffness factor (SF) is the pipe stiffness multiplied by the quantity $0.149r^3$:

$$SF = EI = \frac{F}{\Delta Y} 0.149r^3$$

$$= 0.149r^3 (PS) \qquad (9.19)$$

The stiffness factor or EI is used in the Modified Iowa Formula (Eq. 9.17) to determine approximate field deflections under earth loads. It is the engineer's responsibility to establish the acceptable field deflection limit, and to design the installation accordingly. The manufacturer should be consulted for recommended field installation deflection limits.

Specifications also may require that some types of plastic sewer pipe withstand extreme deflections, such as 40% of the original diameter in a parallel-plate loading test without evidence of splitting, cracking, or breaking. There is no corresponding load requirement. These extreme deflection tests are instantaneous and are intended for production quality control during pipe manufacturing. Although these extreme deflections (observed in ASTM 2412

Table 9-12. Stiffness Requirements for Plastic Sewer Pipe Parallel-Plate Loading*

Material** (1)	ASTM Specification (2)	Nominal diameter, d, in inches (3)	Required stiffness at 5% deflection, in pounds per square inch (4)
ABS Composite	D 2680	8–15	200
ABS Plain	D 2751		
	SDR 23.5	4 & 6	150
	SDR 35	3	50
		4 & 6	45
	SDR 42	8, 10 & 12	20
RPM	D 3262	8–18	Varies (99–17)
		20–108	10
PVC	D 2729 (PVC-12454)	2	59
		3	19
		4	11
		5	9
		6	8
	D 2729 (PVC-13364)	2	74
		3	24
		4	13
		5	12
		6	10
	D 3033		
	SDR 41	6–15	28
	SDR 35	4–15	46
	D 3034		
	SDR 41	6–15	28
	SDR 35	4–15	46

*ASTM 2412
**Other plastic pipe materials are not listed, in that insufficient data is currently available.
Note: 1 in. = 25.4 mm; 1 psi = 6.89 kPa.

testing) are limiting in terms of duration and extent of pipe distortion, they can be important to the design engineer in selecting sewer pipe materials of construction.

b. Design Relationships

Structural design of flexible plastic pipe requires definition of the critical deflection limit for the specific pipe considered. The critical deflection limit for flexible plastic pipe is commonly based on structural performance characteristics and potential modes of failure. The requirements of serviceability, such as effect on carrying capacity, passage of cleaning equipment, velocity and differential deflection at fittings, manholes and joints, also may be considered. Maximum strain in the pipe wall, discussed above, also should be considered.

The design basis is expressed as follows:

$$\text{Design Deflection Limit} = \frac{\text{Critical Deflection Limit}}{\text{Factor of Safety}}$$

In structural design for a flexible sewer pipe system, the design deflection limit should not be more than the maximum long-term deflection anticipated under load. In the determination of projected long-term deflection, the design element considerations are load (soil, superimposed dead load, and superimposed live load); pipe stiffness; and soil stiffness. The primary controlling elements of deflection are the load and the soil stiffness. Sewer pipe stiffness generally is a relatively minor factor by comparison with soil stiffness. The soil stiffness depends on the type of soil and its density as placed in the sewer pipe embedment zone. In consideration of design requirements, the engineer should evaluate the options of different types of flexible sewer pipes, different embedment soils, and different embedment material compaction and placement requirements.

The choice of factor of safety should be influenced by soil characteristics, the degree of compaction likely to be obtained, any available field tests, and practical experience.

c. Loads on Flexible Plastic Pipe

The load carried by a buried flexible pipe in a narrow trench may be calculated using the Marston Formula (Eq. 9.5). A conservative design approach may be used by assuming the dead load carried by a flexible pipe-soil system in any installation to be the prism load. For normal installations the prism load is the maximum load that can be developed.

The load on a projecting flexible sewer pipe is calculated using Eq. 9.6. As before, the load coefficient (C) depends on the projection ratio (p), settlement ratio (r_{sd}), and the ratio of fill height to pipe width (H/B_c), and can be determined from Fig. 9-7. For flexible projecting pipe, the product $r_{sd}p$ is negative or zero. As shown in Fig. 9-7, when the product is zero the load coefficient, C_c, equals H/B_c. Eq. 9.6 then becomes Eq. 9.8, which gives the prism load:

$$W_c = HwB_c$$

d. Field Deflection of Flexible Plastic Pipe

As previously discussed, the pipe stiffness is determined in a standard parallel-plate loading test for an arbitrary deflection of the nominal diameter. This is a measure of the inherent strength of plastic pipe.

The pipe stiffness and soil stiffness as measured by E' must develop sufficient field strength that the deflection of the pipe under load will not exceed the recommended permissible deflection.

Approximate field deflections can be calculated using the Modified Iowa Formula (Eq. 9.17). This formula can be simplified to permit a calculation of approximate deflection based on pipe stiffness as determined by Eq. 9.20:

$$\Delta X = \frac{D_L K_b W_c}{0.149PS + 0.061E'} \qquad (9.20)$$

(Note: If W_c = prism load for flexible pipe, $D_L \sim 1$.)

Values of bedding factors for use in this formula may be found under the "General Methods" subsection of this section.

The solution of Eqs. 9.8 and 9.20 requires that various factors be determined. It is desirable, where possible, to establish anticipated long-term field deflections based on well documented empirical data.

Sample Calculation:

Example 9-14. Determine the type of backfill material and compaction required for a 24-in. RPM pipe with 22 ft of cover in a trench condition. The trench is in ordinary clay with a unit weight of 120 pcf.

$$\text{Assume } B_c = 24.5 \text{ in.} = 2.04 \text{ ft}$$
$$B_d = 4.00 \text{ ft}$$
$$\text{Then } H/B_d = 22/4.0 = 5.5, \ C_d = 2.9$$
$$W_c = C_d \, w \, B_c \, B_d$$
$$= 2.9 \times 120 \times 2.04 \times 4.0$$
$$= 2{,}840 \text{ lb/ft} = 237 \text{ lb/in.}$$

ASTM Specification D3262 for RPM pipe requires that the pipe be able to deflect 10% with no structural damage. Considering 10% a critical deflection limit and applying a factor of safety of 2.0, the design deflection limit is 5%. Then ΔX becomes $0.05 \times 24 = 1.2$ in.

Using Eq. 9.20 and solving for E':

$$\Delta X = \frac{D_L K_b W_c}{0.149PS + 0.061E'}$$

$$E' = \frac{D_L K_b W_c - 0.149PS \, \Delta X}{0.061 \Delta X}$$

Assuming $D_L = 1.5$ and $K_b = 0.102$, according to D3262, $PS = 10$, $E' = 471$ psi

Referring to Tables 9-6 and 9-10, it is noted that Class III soils with moderate compaction or Class IV soils highly compacted will provide an average E' of 1,000 psi. These types of soil and degrees of compaction will be satisfactory. If the average vertical deflection ΔY is assumed to be approximately equal to ΔX, then ΔY may vary ±1 to 2% of the pipe diameter for these average E' values as shown in Table 9-10.

3. Plastic Sewer Pipe Installation

At the present time plastic sewer pipe is not manufactured in a wide range of stiffnesses or wall thicknesses. To enable the pipe to support different heights of fill it is necessary to construct the embedment around the pipe so as to develop the required soil-structure interaction. In general this is measured by the E' value for the installation. Only by increasing the E' value is it possible to support increasingly higher fills and increased loads.

a. Bedding, Haunching and Initial Backfill

The bedding requirements for flexible thermoplastic sewer pipe are given in ASTM D 2321 for Class I, II and III material (Table 9-6). Haunching and initial backfill requirements are also given, including minimum compaction recommendations. By referring to Tables 9-6 and 9-10 it is possible to estimate the E' value obtained for the expected compaction or to determine the compaction requirements needed to develop the E' value which will keep the pipe deflection within allowable limits.

Similarly, bedding and initial backfill requirements for thermosetting reinforced plastic pipe are given in ASTM D 3839, and a guide for estimating both initial and long-term deflection is included in the appendices.

b. Final Backfill

The procedure for installing the final backfill for the remainder of the trench is the same as for any other type of pipe, as described in Section D, "Pipe Bedding and Backfilling."

4. Soil Classification

For different categories of embedment materials, different construction procedures are specified. Soil classifications under the Unified Soil Classification System, including manufactured materials, are grouped into five broad categories according to their ability to develop an interacting sewer pipe-soil system (Table 9-6). The soil classes described below are also given in ASTM D 2321.

For the larger-grained soils (Class I and Class II gravels), compatibility with the existing subgrade and trench side soils should be considered. Particularly for uniformly graded or gap graded materials, the potential exists for the migration of the finer fraction of the existing soils into the embedment materials, with resultant settlement and loss of side support for the sewer pipe. Analytical techniques for assessing this possibility are similar to those for filter blankets and are well covered in the referenced material. The suggested E' values given for each soil class are obtained by reference to Table 9-10.

a. Class I

This class includes angular, 6 to 40-mm (0.25 to 1.5-in.), graded stone, including a number of fill materials that have regional significance, such as coral, slag, cinders, crushed stone, and crushed shells.

Class I material provides the best material for the construction of a stable sewer pipe-soil system. When used for underdraining, Class I material should be placed to the top of the sewer pipe.

Class I material used for haunching and initial backfill, when dumped into place with little or no compaction, will produce an average E' value of 1,000 psi (6.9MPa). Care must be taken to place material under the haunches and in contact with the sides of the pipe. Class I material compacted to 85% Standard Proctor Density or higher has an average E' value of 3,000 psi (21MPa).

b. Class II

This class comprises coarse sands and gravels with maximum particle size of 40 mm (1.5 in.), including variously graded sands and gravels containing small percentages of fines, generally granular and non-cohesive, either wet or dry. Soil types GW, GP, SW and SP are included.

Use of sands of Type SP should be done with caution. Poorly graded fine sands with little material finer than 200-sieve have a tendency to flow when wet.

Class II material used for haunching and initial backfill, when dumped into place with little or no compaction, will produce an average E' value of 200 psi (1.4 MPa). Compacted to 85% Standard Proctor Density, the average E' value is raised to 1,000 psi (7 MPa). Higher E' values are obtained with greater compaction.

c. Class III

This class comprises fine sand and clayey gravels, including fine sands, sand-clay mixtures, and gravel-clay mixtures. Soil types GM, GC, SM and SC are included.

Class III materials dumped in place for haunching and initial backfill will produce an average E' of 100 psi (700 kPa). Careful placement and compaction to 90% Standard Proctor Density will produce an average E' of 1,000 psi (7 MPa).

d. Class IV and V

Class IV materials require special effort for compaction, thus may be suitable for sewer pipe foundation if special care is taken during excavation to provide a uniform, undisturbed trench bottom. Use of Class IV materials for bedding, haunching or initial backfilling is not recommended.

Class V materials present special problems in providing an adequate foundation, as discussed further in Chapter 10. Class V materials should not be used for any part of the sewer pipe envelope.

5. Design of Ductile Iron Sewer Pipes

Five types of bedding are denoted for the installation of ductile iron sewer pipe (DIP) (Fig. 9-22). The design of DIP is based on limiting stresses and deflection, whichever governs.

Stress in the ductile iron pipe wall for gravity flow sewers is caused by flexural stresses from the soil load. Design for flexure is based on the Spangler equation for bending stress (21). This equation utilizes both the bending stress coefficient, K_b, and the deflection coefficient, K_x. Both coefficients are functions of the width of the bedding angle under the pipe.

Ductile iron sewer pipe is also designed for a limiting deflection based on the Modified Iowa Formula. The American National Standard for the Thickness Design of Ductile Iron Pipe (ANSI A21.50) contains design tables for a wide range of soil loads for the five standard laying conditions shown on Fig. 9-22. In using these tables it is not necessary to consider the effect of internal pressure when designing gravity flow sewers. The recommended maximum design deflection of mortar-lined ductile iron pipe is 3% to avoid damage to the mortar lining.

The design ring-bending stress which is recommended by the Ductile Iron Pipe Research Association is 330 MPa (48,000 psi). The following equation may be used to estimate the approximate allowable trench load based on the ring-bending stress:

$$W = \frac{f}{3\left(\dfrac{D}{t}\right)\left(\dfrac{D}{t} - 1\right)\left[K_b - \dfrac{K_x}{\dfrac{8E}{E'\left(\dfrac{D}{t} - 1\right)^3} + 0.732}\right]} \quad (9.21)$$

STRUCTURAL REQUIREMENTS

Laying Condition	Description	E'	Bedding Angle	K_b	K_x
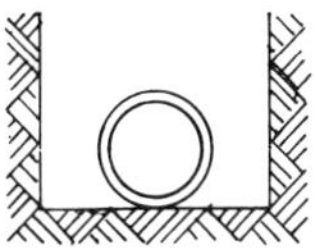 Type 1*	Flat-bottom trench.† Backfill not tamped.	150	30°	0.235	0.108
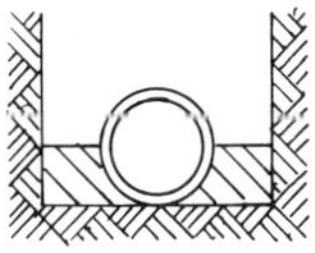 Type 2	Flat-bottom trench.† Backfill lightly tamped** to centerline of sewer pipe.	300	45°	0.210	0.105
 Type 3	Sewer pipe bedded in 4-in. minimum loose soil.†† Backfill lightly tamped** to top of sewer pipe.	400	60°	0.189	0.103
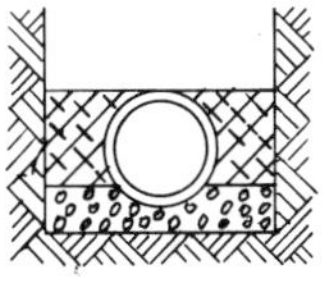 Type 4	Sewer pipe bedded in sand, gravel or crushed stone to depth of 1/8 pipe diameter, 4-in minimum. Backfill lightly compacted** to top of sewer pipe.	500	90°	0.157	0.096
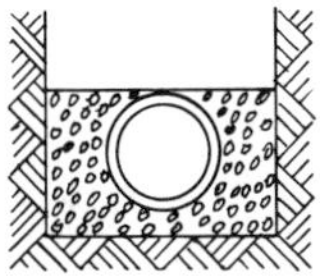 Type 5	Sewer pipe bedded in compacted granular material to centerline of pipe. Carefully compacted** granular or select †† material to top of sewer pipe.	700	150°	0.128	0.085

*Laying condition Type 1 is limited to 16-in. and smaller pipe.
†"Flat-bottom" is defined as undisturbed earth.
††"Loose soil" or "select material" is defined as native soil excavated from the trench, free of rocks, foreign materials and frozen earth.
**These laying conditions can be expected to develop the following backfill densities (Standard Proctor): Types 2 and 3, approximately 70%; Type 4, approximately 75%; Type 5, approximately 85%

Fig. 9-22. Design values for standard laying conditions—ductile iron pipe (in. × 25.4 = mm). Copyright, American Society for Testing and Materials, 1916 Race St., Philadelphia, PA 19103, reprinted with permission.

Table 9-13 Allowances for Casting Tolerances in Ductile Iron Pipe

Size, in millimeters (inches) (1)	Tolerance, in millimeters (inches) (2)
100–200 (4–8)	1.3 (0.05)
250–300 (10–12)	1.5 (0.06)
350–1,050 (14–42)	1.8 (0.07)
1,200 (48)	2.0 (0.08)
1,350 (54)	2.3 (0.09)

where W = total trench load (earth plus live load); f = design bending stress; t = net thickness; D = outside diameter; E = modulus of elasticity; E' = modulus of soil reaction; K_b = bending moment coefficient; and K_x = deflection coefficient.

The following equation is used to estimate the approximate allowable trench load based on deflection:

$$W = \frac{\frac{\Delta Y}{D}}{12 K_x} \left[\frac{8E}{\left(\frac{D}{t} - 1\right)^3} + 0.732 E' \right] \qquad (9.22)$$

where ΔY = design deflection, $0.03D$. Other terms are the same as in the equation for bending stress.

A net thickness is calculated using both the bending stress and deflection equations. The larger of the two thicknesses is selected as the net thickness required for the external load.

The service allowance of 12 mm (0.08 in.) is added to the net thickness to allow for a reduction in thickness because of corrosion, imperfections and the like. A casting tolerance proportional to the diameter is also added. Allowances for casting tolerances are shown in Table 9-13.

6. Design of Corrugated Metal Sewer Pipes

Corrugated metal pipe is manufactured in a variety of gauges, corrugation depths, and corrugation spacings.

The longitudinal seam formed by bolting or riveting curved sheets together for corrugated metal pipe may have to be checked for crushing strength for heavy backfill loads. Tables of seam strengths for various metal gauges and bolt or rivet sizes and spacing can be found in manufacturers' handbooks. Corrugated metal pipe is not used in sanitary sewers without a protective coating which may or may not be asbestos bonded. To improve hydraulic properties the pipe may have a paved invert (25% of circumference) or may have the interior completely paved.

Corrugated metal pipe sewers may be designed for a limiting deflection using the Modified Iowa Formula (Eq. 9.17) or by the manufacturer's handbook figures. A design deflection not greater than 5% is commonly used.

7. Flexible Sewer Pipe Installation

General bedding and backfilling concepts developed for rigid sewer pipe may also be applied to flexible pipe, with some exceptions. The general pipe terminology for installation is shown in Fig. 9-17.

Detailed information on bedding of flexible pipe is contained in the various ANSI and ASTM specifications or manufacturers literature. The design engineer should consult the applicable specification or literature for design information.

a. Bedding, Haunching and Initial Backfill

The bedding requirements for flexible thermoplastic sewer pipe are given in ASTM D 2321 for Class I, II and III material (Table 9-6). Haunching and initial backfill requirements are also given, including minimum compaction recommendations. By referring to Tables 9-10 and 9-12, it is possible to estimate the E' value obtained for the expected compaction or to determine the compaction requirements needed to develop the E' value which will keep the pipe deflection within allowable limits.

Similarly, bedding and initial backfill requirements for thermosetting reinforced plastic pipe are given in ASTM D 3839.

b. Final Backfill

The procedure for installing final backfill for the remainder of the trench is the same as for rigid pipe.

H. REFERENCES

1. Marston, A., and Anderson, A. O., "The Theory of Loads on Pipes in Ditches and Tests of Cement and Clay Drain Tile and Sewer Pipe," Iowa Eng. Exp. Sta., Bull. No. 31 (1913).
2. Marston, A., "The Theory of External Loads on Closed Conduits in the Light of the Latest Experiments". Iowa Eng. Exp. Sta., Bull. No. 96 (1930).
3. Schlick, W. J., "Loads on Pipe in Wide Ditches". Iowa Eng. Exp. Sta., Bull. No. 108 (1932).
4. "Jacked-in-Place Pipe Drainage". *Contractors and Engr.* Monthly, 45 (Mar. 1948).
5. *Jacking Reinforced Concrete Pipe Lines,* Amer. Concrete Pipe Assn., Arlington, Va. (1960).
6. "Report of Test Tunnel," Part I, Vol. 1 and 2, Garrison Dam and Reservoir, Corps of Engineers, U.S. Army.
7. Proctor, R. V., and White, T. L., "Rock Tunneling with Steel Supports," Commercial Shearing and Stamping Co.
8. "Soil Resistance to Moving Pipes and Shafts". *Proc. II Intl. Conf., Soil Mech. and Found. Eng.,* 7, 149 (1948).
9. Von Iterson, F. K. Th., "Earth Pressure in Mining". *Proc. II Intl. Conf., Soil Mech. and Found. Eng.,* 3, 314 (1948).
10. *Vertical Pressure on Culverts Under Wheel Loads on Concrete Pavement Slabs,* Portland Cement Assn., Publ. No. ST-65, Skokie, Ill. (1951).
11. Griffith, J. S., and Keeney, C., "Load Bearing Characteristics of Bedding Materials for Sewer Pipe," *Jour. Water Poll. Control Fed.* 39, 561 (1967).
12. Townsend, M., *Corrugated Metal Pipe Culverts – Structural Design Criteria and Recommended Installation Practices,* Bur. Public Roads, U.S. Govt. Printing Office, Washington, D.C. (1966).
13. Wenzel, T. H., and Parmelee, R. A., "Computer-Aided Structural Analysis and Design of Concrete Pipe", *Concrete Pipe and the Soil-Structural System,* ASTM STP 630, 1977, pp. 105-118.
14. Parmelee, R. A., "A New Design Method for Buried Concrete Pipe" *Concrete Pipe and the Soil-Structural System,* ASTM STP 630, 1977, pp. 105-118.

15. Moser, A. P., Watkins, R. K., and Shupe, O. K., *Design and Performance of PVC Pipes Subjected to External Soil Pressure*, Buried Structure Laboratory, Utah State University, Logan, Utah 84322 (Feb. 1977).
16. *Concrete Culverts and Conduits,"* Portland Cement Association Publication No. EB061.02W, 1975.
17. Schrock, B. J., "Installation of Fiberglass Pipe". *Journal of the Transportation Division, ASCE,* Vol. 104, No. TE6, Proceeding Paper 14175 (Nov. 1978).
18. Watkins, R. K., and Spangler, M. G., "Some Characteristics of the Modulus of Passive Resistance of Soil: A Study in Similitude". *Highway Research Board Proceedings,* Vol. 37, 1958, p. 576-583.
19. Seaman, Donald J., "Trench Backfill Compaction Control Bumpy Streets", *Water and Sewage Works,* Sept. 1979, p. 67.
20. Harell, R. F., and Keeney, Chester, "Loads on Buried Conduit — A Ten-Year Study". *Journal Water Pollution Control Federation,* 48, 1988 (Aug. 1976).
21. Spangler, M. G., "The Structural Design of Flexible Pipe Culverts". Iowa Engineering Experiment Station Bulletin No. 153 (1941).
22. Howard, Amster K., "Modulus of Soil Reaction (E') Values for Buried Flexible Pipe," *Journal of the Geotechnical Engineering Division, ASCE,* Vol. 103, No. GT, Proceeding Paper 12700 (Jan. 1977).
23. Spangler, M. G., "Stresses in Pressure Pipelines and Protective Casing Pipes," *Jour. Str. Div., Am. Soc. Civil Engr.,* Vol. 82, No. ST5, Proc. Paper 1054, Sept., 1956.
24. Jumikis, A. R., "Stress Distribution Tables for Soil Under Concentrated Loads," Engineering Research Publication No. 48, Rutgers University, New Brunswick, NJ, 1969, p. 233.
25. Jumikis, A. R., "Vertical Stress Tables for Uniformly Distributed Loads on Soil," Engineering Research Publication No. 52, Rutgers University, New Brunswick, NJ, 1971, p. 495.
26. Spangler, M. G., and Hennessy, R. L., "A Method of Computing Live Loads Transmitted to Underground Conduits", Proceedings, 26th Annual Meeting, Highway Research Board, 1946, page 179.
27. American Association of State Highway and Transportation Officials, "Standard Specifications for Highway Bridges," 12th Edition.
28. Sikora, E. J., "Load Factors and Non-Destructive Testing of Clay Pipe," *Journal Water Pollution Control Federation,* December 1980, page 2964.
29. Taylor, R. K., "Final Report on Induced Trench Method of Culvert Installation", Project IHR-77, State of Illinois, Department of Public Works and Buildings, Division of Highways, Springfield, Illinois, November 1971.
30. American Railway Engineering Association, "Manual for Railway Engineering," 1981-82, pp. 1-4-26 and 8-10-10, Washington, DC.

CHAPTER 10

CONSTRUCTION CONTRACT DOCUMENTS

A. INTRODUCTION

The purpose of the contract documents is to portray clearly by words and drawings the nature and extent of the work to be performed, the conditions known or anticpated under which the work is to be executed, and the basis for payment. Most sewer construction projects are accomplished by contracts entered into between an owner and a construction contractor. The contract documents, consisting of the bidding requirements, bid forms, contract forms, conditions of the contract, specifications, plans and addenda, constitute the construction contract.

The contract drawings and the specifications together define the work to be done by the contractor under the terms of the construction agreement. These documents are complementary; what is called for by one is to be executed as if called for by both.

The contract documents establish the legal relationship between the owner and the contractor, as well as the duties and responsibilities of the engineer. It follows that the engineer's potential liability is influenced by the contract documents although the engineer is not a party to the construction contract.

Experience in the courts over many years has revealed a number of sensitive areas of potential liability which may not be covered in the owner's and the engineer's in-house documents. Among these are (1) means and methods of construction, (2) right to stop the work, (3) safety, (4) insurance, (5) supervision, and (6) indemnification. In general terms, the contractor is responsible for means and methods of construction, safety and supervision of workmen; only the owner has the right to stop the work, and his insurance adviser makes final decisions on insurance provisions. Subject to limitations of State law, the engineer should be included with the owner as a party indemnified.

These provisions and many others are covered in an adequate manner in the standard documents of the Engineers Joint Contract Documents Committee (1). These standard documents are recommended as the basis for developing the contract documents. They are further discussed in Section C of this chapter. Of course, good legal advice is prudent in all contractual matters.

Prior to bidding, the project usually requires approval by the regulatory agencies. When approved, a permit to construct the project may be issued. After the project has been approved and the contract documents are in final form, bids are solicited.

For ease in bidding and administration, frequently the work is divided into various items, with either unit or lump sum prices received for each item of work. The contract documents must clearly describe and limit these items to obviate all possible confusion in the mind of the bidder with regard to methods of measurement and payment. The subdivision of the work often is based on local customs, the customs and conventions of the design engineer or the specific requirement of the owner.

Lump sum bids have been applied most generally to special structures which are completely defined and not subject to alteration or quantity changes during construction. A schedule of unit adjustment prices may be included in the proposal to provide a basis for payment in the event that changes are necessary in lump sum bid items. These unit adjustment prices are frequently stipulated by the engineer to assure uniform comparison of proposals. Stipulated prices should always be fair and close to bid price experience for similar work.

Unit price bids have been used most generally where quantities of work are likely to be variable and adjustment found necessary during construction. Linear feet (meters) of sewer, numbers of manholes and cubic yards (cubic meters) of rock excavation or concrete cradle are examples of such unit price items.

Lump sum bids may be taken for an entire sewer construction contract where the contract documents define the work with sufficient completeness to permit the bidder to make an accurate determination of the quantities of work. Such contracts may contain unit adjustment prices for items of work, such as rock excavation, piles, additional excavation, selected fill material, and sheeting requirements which cannot be determined precisely beforehand. The unit adjustment price should be stipulated in the proposal or an appropriate quantity of the unit price work may be included for comparison of bids. The administration of the project, provided extensive changes are not made during construction, is simplified in the lump sum type of contract.

A final caveat — administration of the contract is controlled by the contract documents which must include applicable administrative contract provisions required by federal, state and local governments. The more the contract documents put the engineer in "control" of the contractor's activities, rather than of the results of the contractor's work, the greater the duties and legal obligation imposed on the engineer.

B. CONTRACT DRAWINGS

1. Purpose

The purpose of the contract drawings (plans) is to convey graphically the work to be done, first to the owner, then to project reviewers, to the bidders and later to the construction observers/inspectors/engineer and the contractor. All information which can best be conveyed graphically, including configurations, dimensions and notes, should be shown on the contract drawings. Lengthy word descriptions are best included only in the specifications, and need not be repeated on the contract drawings.

2. Field Data

A survey and investigation of the route of the sewer is required to obtain information as to the existing topography, underground utilities and property boundaries to be shown on the contract drawings. The route may be mapped from data obtained by conventional ground surveys or by aerial photogrammetry and should be checked against any existing plans and records if they are available. Survey work is discussed in some detail in Chapter 2.

When the location of the sewer has been well defined by preliminary studies, it may be possible to run the ground survey baseline directly on the centerline of the proposed alignment. This procedure will facilitate office plotting of field data and later simplify stakeout of the sewer for construction. If the actual alignment is not established by field surveys, baselines or reference marks must be established in the field.

3. Preparation

Contract drawings generally are prepared on a translucent medium in pencil or ink to facilitate reproduction. Whether pencil or ink is used and the type of medium are matters of local practice. When the drawings are to be used for "permanent" records, either ink is used or black line prints on mylar or an equivalent medium are made from pencil drawings. Printed plan and profile sheets are available or may be specially printed and titled to reduce drafting time. The plan view is usually drawn on the top half of the sheet and the profile is plotted directly beneath it on the bottom half, facilitating coordination of the two views. The profile grid is usually printed on the back of the sheet so as not to be erased when corrections are made to the profile.

Topography for the sewer plan is plotted either from the ground survey field notes or from the aerial photography, resulting in a strip map showing relevant existing facilities and ground surface features along the path of the proposed sewer. This topography may be transferred directly to the sewer plan/profile sheets from which prints can be made for use in sewer design; when aerial photogrammetry is used, the topography may be plotted directly on the aerial photographs and the resulting photomap may be reproduced in the configuration of a strip plan for use in design. The plotted topography and record data obtained from utility companies as to underground utilities are plotted on the plan. The plan showing existing conditions is completed by plotting property and easement boundaries from tax map information, deed descriptions, property maps of individual parcels and physical evidence obtained by field survey. Relevant basement elevations, inverts of existing utilities or other structures that may effect the work may be shown.

The proposed sewer alignment is developed on prints of the plan/profile sheets or photomaps showing existing conditions. A profile of the ground surface along the proposed sewer alignment is drawn and proposed sewer invert elevations are determined and plotted on the profile. The horizontal and vertical alignments of the proposed sewer are transferred from the work prints to reproducible plan/profile sheets. Sections and details of the proposed sewer and appurtenances are drawn and traced on reproducible sheets for incorporation into the contract drawings. When applicable, proposed temporary and permanent easement boundaries are determined and drawn on the sewer plans. Contract limit lines are added to complete the preparation of the sewer plans.

The scale of the sewer plans should be large enough to show all of the necessary surface and subsurface information without excessive crowding. A horizontal scale of 1:500 to 1:1000 [40 to 100 ft to the inch (5.0 to 10 m to the centimeter)] is suitable for most sewer plans; in extremely crowded urban areas 1:250 [20 ft to the inch (2.5 m to the centimeter)] should be considered. In such areas, large-scale plans of street intersections are quite useful. Generally a satisfactory vertical scale for the profile is 1:100 [10 ft to the inch (1.0 m to the centimeter)] or 1:50 [5 ft to the inch (0.5 m to the centimeter)] for relatively flat terrain. Larger scales are used for sections and details. Contract drawings are sometimes reduced to approximately one-half scale and issued to bidders in this size for convenience. This practice requires careful preparation of the full-size drawings to produce clear and readable reductions. Reduced size plans should contain a note stating the magnitude of size reduction (if it is an exact reduction, such as half-size) and should always have a graphical scale.

When plans and profiles are drawn using metric dimensions, stations each will be 100 m; horizontal dimensions should only be shown to 0.01 m. Bench mark elevations will be given to 1 mm (i.e., three places of decimals). However, it will generally be appropriate to show sewer inverts, etc., only to two decimal places. Nominal sewer sizes should always be shown in millimeters; for soft conversion of diameters in inches, it is generally acceptable to multiply by 25 (instead of the exact 25.40) to get the nominal diameter in millimeters.

Lettering on contract drawings falls into three general categories: Labeling and dimensioning, notes, and titles. Labeling and dimensioning of existing conditions is generally done by freehand lettering with a minimum letter size of No. 4. If the drawings are later to be reduced in size, a minimum letter size of No. 5 should be considered. Proposed facilities are labeled and dimensioned using lettering producing the same size or larger lettering as that used for existing conditions. Labels and dimensions associated with proposed facilities should stand out from those for existing facilities. Notes should be lettered in the size used for labels and dimensions for proposed facilities. Titles should consist of mechanically produced letters larger than those used for labeling and dimensioning or notes.

Contract drawings should be prepared carefully in a neat, legible fashion. Hastily produced, sloppy drawings can lead to mistakes during construction which might be blamed on the condition of the drawings.

4. Contents

a. Arrangement

The most logical arrangment for a set of contract drawings develops the project from general views to more specific views, and finally to more minute details. The subsections below are arranged to follow this generally accepted order of plan presentation.

b. Title Sheet

The title sheet should identify the project by presenting the following information:
- Project name,
- Contract number,
- Federal or state agency project number (if applicable),
- Owner's name,
- Owner's officials, key people or dignitaries,
- Design engineer's name,
- Engineer's project number,
- Plan set number (for distribution records),
- Professional engineer's seal and signature.

c. Title Blocks

Each sheet except the title sheet should have a title block containing the following:
- Sheet title,
- Project name,
- Federal or state agency project number (if applicable),
- Owner's name,
- Design engineer's name,
- Sheet number,

- Engineer's project number,
- Scale,
- Date,
- Designer, drafter and checker identification,
- Revisions block,.
- Sign-off by owner's chief engineer or district superintendent.

d. Index/Legend

Contract drawings should contain an index which lists all of the drawings in the set by title and drawing number in order of presentation. It also is useful to include a sheet index drawn on the general plan map to identify the sheets which show the details for each length of proposed sewer shown on the general plan. These indices should be located on the drawing following the title sheet. A legend showing a set of symbols for elements of topographic abbreviations and the various items of the sewage works indicated in the sewer plans should be included on the index sheet. An example of a legend for sewer plans is shown in Fig. 10-1.

e. Location Map

There should be a general location map showing the location of all work in the contract in relation to the community, either on the title sheet or on the index/legend sheet. This location map also may be used as an index map as outlined in the preceding paragraph.

f. General Notes

Notes which pertain to more than one drawing should be presented on the index/legend sheet. An example is a note warning that the location and sizes of existing underground utilities shown on the contract drawings are only approximate and that it is the responsibility of the contractor to confirm or locate all underground utilities in the area of his work. Each sheet should contain a note warning against the unauthorized alteration or addition to the documents.

g. Subsoil Information

Whether or not the locations of soil borings made during the design phase of a project and the boring logs should be included in the contract documents is a decision that should be made only after proper legal advice and consideration. In any event, whatever subsurface information has been obtained should be made available to bidders.

Drawings and specifications should indicate where special construction is required because of known unfavorable subsoil conditions. Neither the owner nor the engineer should guarantee subsoil conditions as a known element of the contract agreement.

h. Survey Control and Data

Survey control information may be shown on the sewer plan/profile sheets or on a general plan on a separate sheet. Baseline bearings and distances should be included with reference ties to permanent physical features. Vertical control points, or benchmarks, should be indicated, and the datum plane used for determining these elevations must be defined. A note indicating the dates of the ground survey and aerial photography should be included.

228 GRAVITY SANITARY SEWER

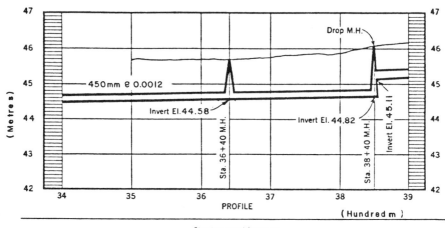

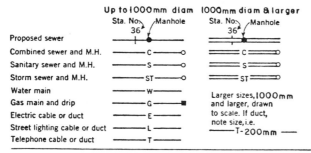

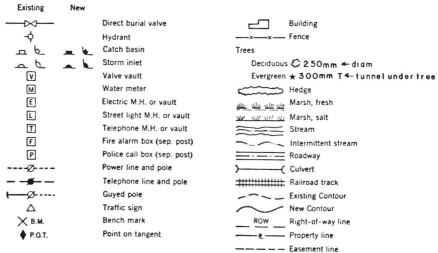

Fig. 10-1. Typical legend for sewer plans (m × 3.3 = ft; mm × 0.039 = in.).

i. Sewer Plans

A continuous strip map, drawn directly above the profile, to indicate the plan locations of all work in relation to surface topography and existing facilities is an integral part of each sewer plan and profile drawing. Underground and overhead utilities along, across or near the proposed construction route also should be shown.

The width of the strip map should be such that only topography which directly affects construction or access to the work is indicated.

Plans for sewers to be constructed within easements on private property should show survey baseline and sewer alignment data. Widths of proposed temporary and permanent easements should be dimensioned.

Sewer plans always should be oriented so that the flow in the sewer is from right to left on the sheet. Each sewer plan should include a north arrow consistent with this arrangement. Stationing should be upgrade from left to right, generally along the sewer centerline. Survey baseline stationing also may be given on the plans, but it should not be substituted for stationing along the centerline of the sewer.

Stationing indicated on construction drawings for location of manholes and wye-branches or house connections is to be considered approximate only and should be so noted. Record drawings after construction must, however, give accurate locations of all features of a completed sewer system. Locations of junction structures must be held firm as given on construction drawings. Match lines should be used and should be easily identifiable. Special construction requirements, such as sheeting to be left in place, should be shown on the drawings. Where interference with other structures is known to exist, explanatory cross-sections and notes should be included. Such cross-sections, often enlarged in scale, should be identified as to specific location and, if practicable, should be placed on the plan/profile drawing near where the section is cut.

A plan and profile for a sewer to be constructed in an extremely congested urban area is given in Fig. 10-2, and for a representative residential area in Fig. 10-3. For congested areas it is desirable to include strip maps of existing conditions and proposed sewer facilities on the same plan/profile sheet. However, when this is not possible, the existing conditions can be placed on sheets separate from the proposed sewer plan/profile sheets.

j. Sewer Profile

Contract drawings should include a continuous profile of all sewer runs indicating centerline ground surface and sewer elevations and grades. The profile is also a convenient place to show the size, slope and type of pipe, the limits of each size, pipe strength or type, the locations of special structures and appurtenances, and crossing utilities and drainage pipes. The profile should be located immediately under the plan for ready reference. Stationing shown on the plan should be repeated on the profile.

k. Sewer Sections

When sewers consist of pipes of commonly known or specified dimensions, materials, or shapes, no sewer sections need to be shown. For cast-in-place concrete sections, complete dimensions with all reinforcement steel shown should be included in the drawings.

l. Sewer Details

Separate sheets of sewer details normally follow the plan/profile sheets.

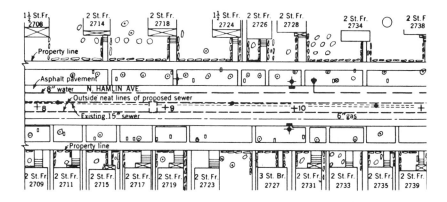

(a) Plan of existing

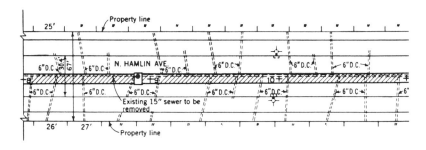

(b) Sewer

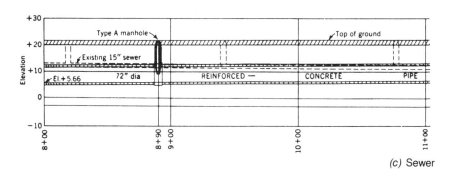

(c) Sewer

Fig. 10-2. Plan and profile for major sewer in congested city street (ft × 0.3 = m; in. × 25.4 = mm).

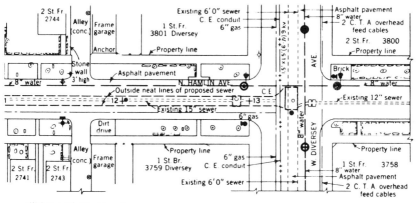

utilities and topography

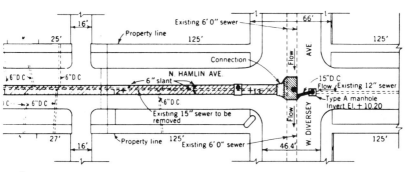

plan

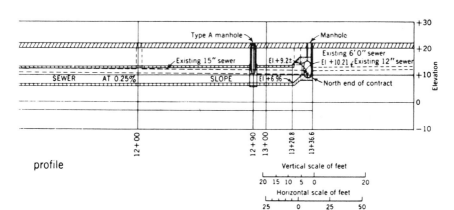

profile

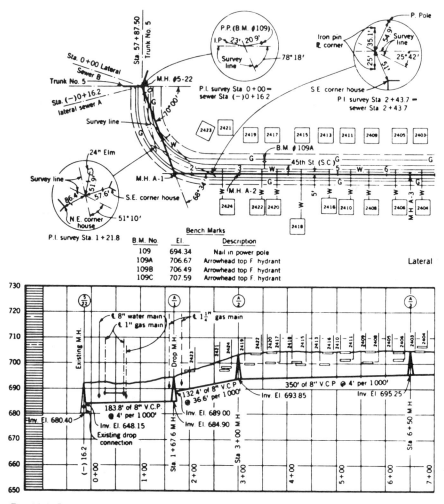

Fig. 10-3. Plan and profile for lateral sewer in residential street (ft×0.3=m; in.×25.4=mm).

The following details, when applicable, should be included:
- Trenching and backfilling — payment limits including those for rock excavation and types of backfill materials.
- Pipe bedding and cushion — dimensions, material types and payment limits.
- House lateral connections — type and arrangement of fitting and minimum pipe grade.
- Special connections — type and configuration of fittings and dimensions.
- Manholes — foundation, base, barrel, top slab, frame and cover, and invert details.
- Sewer/watermain crossings — separation requirements.
- Stream, highway or railroad crossings — casing, inverted siphon, encasement or other related details.

CONSTRUCTION CONTRACT DOCUMENTS

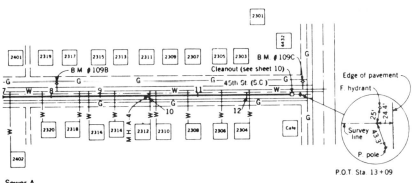

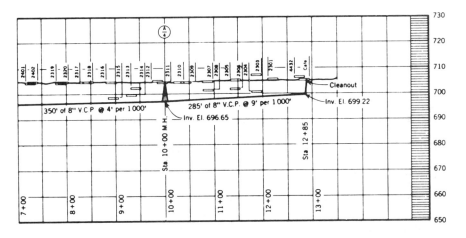

Many of these details find repeated use in sewer projects. It is helpful to develop standard details of these items which may be reproduced for repeated use in sewer contracts.

m. Special Details

Details which do not pertain directly to the sewer piping and which are not covered by standard details should be given on sheets entitled "Special Details." The following would be included:
- Special structures — full details so that the finished work is structurally sound and hydraulically correct.
- Special castings — sufficient details for the manufacturers to prepare shop drawings. Standard casting items, such as manhole frames, covers and manhole steps, will be identified by reference to a manufacturer's catalog number in the specifications.

- Restoration items — complete details for pavement, sidewalk and curb repairs.

5. Record Drawings

During construction of the sewer project, the contractor or the engineer should measure and record the locations of all wyes, stubs for future connections, and other buried facilities which may have to be located in the future. All construction changes from the original plans, rock profiles, and other special classes of excavated material also should be recorded by the contractor or the engineer.

Contract drawings should be revised to indicate this field information after the project is completed and a notation such as "Revised According to Field Construction Records" or "Record Drawing" should be made on each sheet. Record sets of such revised drawings should become a part of the owner's permanent sewer records.

C. PROJECT MANUAL

1. Introduction

The bound volume containing the bidding documents, the agreement forms, the conditions of the contract and the specifications, preferably is termed the "Project Manual". The commonly used title "Specifications" is misleading in that the bound volume contains far more than the material and workmanship requirements for the project, which is the definition of the term "Specifications".

The introduction to this chapter (Section A) recommended that the standard documents of the Engineers Joint Contract Documents Committee (1,4) be used as a basis for developing contract documents. Due to the increased legal exposure of the engineer and owner, the adoption of these standard documents, to benefit from mutual legal experience, should be considered.

These documents are closely related and coordinated; changes in one may require changes in one or more of the others. Further, the standard documents, as well as in-house documents, should be reviewed by the attorneys for both the owner and the engineer for each project due to their legal consequences.

Standard documents generally provide acceptable documents, and through widespread usage are better understood and less subject to misinterpretation. Further, they offer the user language which has been tested by the courts.

While there is a saving of time by utilizing standard documents in preparation of contract documents, one should guard against irrelevant or contradictory requirements within the entire contract document, particularly between standard and specific portions of the contract. Again, good legal advice is essential.

2. Purpose

Documents contained in the project manual set forth the details of the contractual agreement between the contractor and owner. They describe the work to be done, complementing the information provided on the drawings, and establish the method of payment. They also set forth the details for the performance of the work including necessary time schedules and requirements for insurance, permits, licenses, and other special procedures.

Documents contained in the project manual must be clear, concise and complete. All portions should be written to avoid ambiguity in interpretation. Specifications should be understood easily and should be devoid of unnecessary words and phrases, yet they must completely outline the requirements of the project. Reference to standards such as those of ASTM can be used to reduce the bulk of the specifications without detracting from completeness. A high degree of writing skill and knowledge of standards are needed to produce a good set of documents.

3. Arrangement

The arrangement of the contents of the project manual varies, depending to an extent on the requirements of the owner and the practices of the design engineer. Furthermore, arrangement and division of the contents frequently are subject to local legal requirements.

Many government agencies, private owners and engineers have adopted the practices established by the Construction Specifications Institute (CSI) (2,5). As a means of standardizing the order of the documents and the location of contract subject matter within the project manual, and thereby improve communications among members of the construction team, it is recommended that the practices embodied in the CSI Manual of Practice be considered in the preparation of the project manual. The Engineers Joint Contracts Documents Committee's "Uniform Location of Subject Matter" (1f) is recommended as a guide in the preparation of project manuals to further aid in accomplishing improved communications.

Preferably all parts of the project manual are bound in a single volume. However, for extensive programs of sewer construction, a "Standard Project Manual" may be developed, to be bound separately and incorporated by reference in manuals for individual but related projects.

The project manual can be divided most logically into parts that each define a phase or function in the overall administration and performance of the contract. Each part satisfies a definite requirement. Information properly included in one portion generally should not be repeated in others.

The assembled volume should be prefaced with a cover page, title page and table of contents. A convenient arrangement which organizes the parts in their logical order of use is as follows:
 (a) Addenda
 (b) Bidding Requirements
 (c) Contract Forms
 (d) Conditions of the Contract
 (e) Specifications

General notes on these sections may be found on the pages following, with details covered in the included check list.

4. Addenda

Addenda are issued during the bidding period to correct errors and omissions in the contract documents, to clarify questions raised by bidders and to issue additions and deletions to the documents. Changes to both graphic and written documents may be accomplished by an addendum.

Procedures for issuing addenda are described in the instructions to bidders, and space for the bidder to acknowledge their receipt is provided in the bid form. Addenda must be issued sufficiently in advance of the bid opening to give bidders time to actually use it in the preparation of bids.

5. Bidding Requirements

Bidding requirements should be clearly set forth in the first part of the project manual. This part covers all requirements, instruction, and forms pertaining to the submission of bids. In respect to the use of the standard documents, the Engineers Joint Contract Documents Committee excludes bidding documents from their definition of Contract Documents; experience favors the practice of many engineers who include the bidding documents in the project manual and as part of the Contract Documents. The bidding requirements include the following items.

a. Invitation to Bid

This document is sometimes called the "Advertisement for Bids," or the "Notice to Contractors." It should be brief and simple, containing only the information essential to permit a prospective bidder to determine whether the work is in his line, whether he has the capacity to perform, whether he satisfies the pre-qualification requirements, whether he will have time to prepare a bid, and how to obtain bid documents.

b. Instructions to Bidders

The instructions to bidders furnish prospective bidders with detailed information and requirements for preparing and submitting bids. Included here are bidder's responsibilities and obligations; the method of preparation and submission of proposals; the manner in which bids will be canvassed, the successful bidder selected and the contract executed; and other general information regarding the bid-award procedure.

c. Bid Form

The purpose of a stipulated form is to ensure systematic submittal of pertinent data by all bidders in a form convenient for comparison. It must be so worded and prepared that all bidders will be submitting prices on a uniform basis.

The bid form is addressed to the owner of the proposed work and is to be signed by the bidder. The project for which the bid is submitted must be identified. The bid form contains spaces for insertion of unit or lump sum prices and may also contain spaces for bidder's extensions for each item and a total bid price; however, it must be stated that bidder's extensions and total bid price are unofficial and subject to verification by the owner. Bid prices are commonly stated in both words and figures with the word description governing in case of a discrepancy. The bid form may provide for taking bids on alternative materials or methods of executing portions of the work. It may provide for combination bids on several contracts in the project. The bases for considering alternatives and combinations must be described in the instructions to bidders and set forth in the bid form. An informal comparison of bids first may be made, based on the totals given in the submitted bid form. The formal bid comparison must be made after the extension of unit prices and totals of contract items have been checked and determined by the owner to be correct.

The completion date or time is generally set by the owner so that all bidders are submitting prices on the same time basis and the only variable is price. However, on some projects, the time allowed for construction or completion may be set by the bidder as part of his bid. The latter practice is generally not recommended due to the complexities created in evaluating and comparing bids. If this course is chosen, criteria for evaluating the time for completion must be established in the instructions to bidders.

The bid form contains spaces for the bidders to acknowledge the receipt of addenda issued during the bidding period.

Statements must also be included to the effect that the bidder has received or examined all documents pertaining to the project, that the requirements of addenda were taken into consideration in rendering the bid, and that all documents and the site have been examined.

6. Contract Forms

The contract forms constitute the legal framework for the agreement between the contractor and owner. Once executed, each party is bound to fulfill its responsibilities as described in the contract documents. They comprise the form of contract, surety bonds, and special forms.

a. Form of Contract

The form of contract is regulated by the laws of the local jurisdiction and state in which it is executed. The contract must, however, cover all items of work included in the bid except those that have been eliminated as alternative items. It also must bind the contracting parties to conformity with the provisions of all the contract documents.

b. Surety Bonds

Acceptable bond forms should be included in the related documents and signing of the form of contract is contingent on prior receipt of the executed bonds.

Bonds must be executed by a financially responsible and acceptable surety company (3). General practice is to require performance and payment bonds and that each be in the amount of the contract bid price. Maintenance or guarantee bonds may also be required.

c. Special Forms

Include here any special provisions which may apply.

7. Conditions of the Contract

This portion of the contract documents is concerned with the administrative and legal relationships, rights and responsibilities between the owner, the owner's representatives, the contractor, subcontractors, the public and other contractors. Conditions of the contract should contain instructions on how to implement the provisions of the contract. They should not include detailed specifications for materials or workmanship nor work-related administrative and non-legal matters.

The conditions of the contract consist of two parts: the general conditions and the supplementary conditions.

Standard general conditions have been developed by professional societies, as well as by government agencies and private owners. Consultants frequently have in-house standards for projects of their design.

Standard general conditions usually require revisions, deletions or additions to suit the needs of a particular project. These changes are incorporated into the supplementary conditions.

The carefully chosen language contained in the standard general conditions should be modified only when necessary for a specific project as may be required by the locale of the project, the requirements of the owner or the complexity of the project. All modifications must be coordinated with the other documents to avoid the creation of contradictory requirements within the contract documents and should be reviewed by legal counsel.

Many currently published standard general conditions contain articles of a non-legal nature, such as submittals. With the general acceptance of the CSI Format(2) and its proper utilization, non-legal matters more appropriately should be dealt with in "Division 1, General Requirements" of the specifications.

8. Specifications

Specifications cover qualitative requirements for materials, equipment and workmanship as well as administrative, work-related requirements. The two types of provisions should be kept separate.

The CSI Format(2) provides a 16 "division" framework for the development of project specifications. Each division consists of a number of related "sections." Division titles are fixed; section titles are left to the discretion of the specifier. Many divisions are not applicable to sewer work and would be deleted, with the remaining divisions utilized to define project requirements (see included check list).

a. General Requirements (CSI Division 1)

Individual characteristics regarding conditions of the work, procedures, access to the site, coordination with other contractors, scheduling, facilities available, and other non-legal, work-related and administrative details which are unique to the particular contract are placed properly in the general requirements.

CSI practices would require that certain published standard general conditions be modified to delete administrative and work-related provisions contained in these documents and that these provisions be dealt with in the general requirements. Among these provisions are shop drawings, substitutions, project record documents and schedules.

The decision as to which document should be used to deal with certain matters may best be resolved by referring to the Uniform Location of Subject Matter (1f).

b. Material and Workmanship Specifications (CSI Divisions 2 to 16)

Specific details regarding materials or workmanship applicable to the project may be written especially for the contract, or general specifications called "Standard Specifications" may be prepared which are intended to apply to many contracts. In addition, government and private organizations in many areas have developed standard specifications for sewer construction which reflect local practices. These standards should be reviewed to obtain an understanding of common practices in a particular locale.

Commonly, materials are specified by reference to specifications of the American Society for Testing and Materials (ASTM), American National Standards Institute (ANSI), American Concrete Institute (ACI), American Water Works Association (AWWA) and other similar organizations.

Standards of workmanship should be described in specific terms when feasible, but specification of construction means and methods and safety always should be avoided. Nonetheless, parameters and limits often must be specified to assure that construction procedures will be consistent with design intent.

For a major sewer project, the check list covers the majority of material and workmanship specifications which may be required.

9. Modifications

Modifications consist of written amendments to the contract documents signed by both parties to the agreement, change orders signed by the owner and field orders issued by the engineer. Modifications are issued after the execution of the agreement.

Modifications serve to clarify, revise, delete from or add to existing contract documents. They serve to rectify errors, omissions and discrepancies, and to institute design changes requested by the owner or made necessary by unanticipated conditions encountered in carrying out the work.

The procedures for effecting modifications must be clearly spelled out in the contract documents.

10. Supplementary Information

Frequently special studies such as soil analyses and soil borings are made during the design phase of a project. The results of such soil investigations and samplings with soil classifications should be made available to the prospective bidders during the bidding period.

All available information pertinent to the work, especially with reference to subsurface conditions, should be made available to the bidders for examination. Bidders should be obliged to make their own interpretations of the subsoil information.

Arrangements should be made with the owner to permit bidders to inspect the soil samples and to make such additional soil borings as they may deem necessary prior to the bid date.

Competent legal advice should be obtained in deciding whether or not such information should be made a part of the contract documents.

11. Standard Specifications

General specifications for workmanship and materials are intended to provide detailed descriptions of acceptable materials and performance standards which can be applied to all sewer contracts in a given jurisdiction. The description of acceptable construction procedures should be avoided in standard specifications. Any time that construction procedures are specified, care should be exercised to prevent the substitution of fixed concepts for the contractor's initiative. In every case procedures must achieve, safely, specified final results.

General specifications usually are aimed at more than any one specific contract. They may be used on a group of similar contracts or even for larger groups of dissimilar contracts.

A supplement to the standard specifications may be written to include special requirements modifying the standard specifications for a particular contract. In this regard, care must be taken in using standard specifications so that the work involved in writing supplements is not greater than the work that would be required to write completely new specifications for the contract. The use of computer-produced specifications facilitates converting standard specifications into project-specific ones.

It may be necessary to write a long supplementary specification due to the fact that the standard specifications are promulgated into law by a governing body. The engineer or owner cannot supersede this document except by supplementing the unwanted provisions in the governing body's standard specifications.

12. Project Manual Check List

The project manual must cover all the legal-contractual and specifications requirements for the contract. The CSI Format (2) provides an excellent basis for a project manual check list. The check list delineates but does not classify these items in detail. Titles given in the listing refer to CSI broadscope titles; reference to the CSI Format will assist in a more complete, narrow-scope title breakdown.

As a guide for determining completeness of construction documents, the following check list of subjects is offered:

(a) **Bidding Requirements**
- Invitation to Bid:
    (1) Identification of owner or contracting agency.
    (2) Name of project, contract number, or other positive means of identification.
    (3) Time and place for receipt and opening of bids.
    (4) Brief description of work to be performed.
    (5) When and where contract documents may be examined.
    (6) When and where contract documents may be obtained; deposits and refunds therefor.
    (7) Amount and character of any required bid security.
    (8) Reference to further instructions and legal requirements contained in the related documents.
    (9) Statement of owner's right to reject any or all bids.
    (10) Contractor's registration requirements.
    (11) Bidder's pre-qualification, if required.
    (12) Reference to special federal or state aid financing requirements.

- Instruction to Bidders:
    (1) Instructions in regard to bid form including: Method of preparing, signing, and submitting same; instructions on alternatives or options; data and formal documents to accompany bids; etc.
    (2) Bid security requirements and conditions regarding return, retention and forfeiture.
    (3) Requirements for bidders to examine the documents and the site of the work.
    (4) Use of stated quantities in unit price contracts.
    (5) Withdrawals or modification of bid after submittal.
    (6) Rejection of bids and disqualification of bidders.
    (7) Evaluation of bids.
    (8) Award and execution of contract.
    (9) Failure of bidder to execute contract.
    (10) Instructions pertaining to subcontractors.
    (11) Instructions relative to resolution of ambiguities and discrepancies during bid period.
    (12) Contract bonding requirements.
    (13) Governing laws and regulations.

- Bid Form:
    (1) Identification of Contract.
    (2) Acknowledgement of receipt of addenda.
    (3) Bid prices (lump sum or unit prices).

CONSTRUCTION CONTRACT DOCUMENTS

(4) Construction time or completion date.
(5) Amount of liquidated damages.
(6) Financial statement.
(7) Experience and equipment statements.
(8) Subcontractor listing.
(9) Contractor's statement of ownership.
(10) Contractor's signature and seal.
(11) Noncollusion affidavit.
(12) Consent of surety.

(b) **Contract Forms**
- Form of Agreement:
    (1) Identification of principal parties.
    (2) Date of execution.
    (3) Project description and identification.
    (4) Contract amount with reference to the contractor's bid.
    (5) Contract time.
    (6) Liquidated damage clause, if any.
    (7) Progress payment provisions (may be covered in Conditions of the Contract or Division 1).
    (8) List of documents comprising the contract (may be covered in Conditions of the Contract or Division 1).
    (9) Authentication with signatures and seals.

- Bonds
    (1) Performance Bond.
    (2) Labor and Material Payment Bonds.
    (3) Maintenance and Guarantee Bonds (if required).

- Special Forms

(c) **Conditions of the Contract**
- General Conditions.

- Supplementary Conditions.

(d) **Specifications**
- General Requirements (CSI Division 1)
    (1) Summary of work.
    (2) Alternatives.
    (3) Measurement and payment (many offices include this in the particular work item).
    (4) Project meetings.
    (5) Submittals.
    (6) Quality control.
    (7) Temporary facilities and controls (protection).
    (8) Material and equipment.
    (9) Project closeout.

- Site Work (CSI Division 2)
    (1) Existing Utilities and Underground Structures:
        (a) Protection
        (b) Relocation
    (2) Clearing:

(a) Tree removal.
(b) Pavement removal.
(3) Earthwork:
    (a) Excavating, backfill and compacting.
    (b) Limits on width of trench.
    (c) Spoil placement.
    (d) Preparation of trench bottom.
    (e) Pipe bedding.
(4) Pipe Boring and Jacking.
(5) Tunneling:
    (a) Excavating.
    (b) Casing installation.
(6) Sheeting and shoring.
(7) Rock excavation:
    (a) Definition of rock.
    (b) Excavation.
    (c) Blasting limitations and control.
(8) Site drainage.
(9) Paving and surfacing:
    (a) Streets and roadways.
    (b) Sidewalks.
(10) Highways and railroad crossings.
(11) Piping materials and jointing.
(12) Manholes and appurtenances.
(13) Pipe laying:
    (a) Control of alignment.
    (b) Control of grade.
(14) Service connections.
(15) Connections to existing sewers.
(16) Connections between different pipe materials.
(17) Concrete encasement or cradle.
(18) Sewer paralleling water main.
(19) Sewer crossing water main.
(20) Repair of damaged utility services.
(21) Acceptance tests:
    (a) Infiltration.
    (b) Exfiltration.
    (c) Smoke.
    (d) Air.

- Concrete (CSI Division 3)
    (1) Forms.
    (2) Concrete reinforcement.
    (3) Cast-in-place concrete.
    (4) Concrete curing.

- Metals (CSI Division 5)
    (1) Metal fabrications (manhole steps and covers, castings).

- Finishes (CSI Division 9)
    (1) Painting.
    (2) Waterproofing.

D. REFERENCES

1. Engineers' Joint Contract Documents Committee, (including as member organizations: American Society of Civil Engineers, National Society of Professional Engineers, and American Consulting Engineers Council) and The Construction Specifications Institute:
   a. "Standard General Conditions of the Construction Contract," Document Number 1910-8.
   b. "Standard Form of Agreement Between Owner and Contractor"
      — "On the Basis of a Stipulated Price," Document Number 1910-8-A-1.
      — "On the Basis of Cost-Plus," Document Number 1910-8-A-2.
   c. "Standard Form of Instruction to Bidders," Document Number 1910-12.
   d. "Guide to the Preparation of Supplementary Conditions," Document Number 1910-17.
   e. "Suggested Bid Form and Commentary For Use," Document Number 1910-18.
   f. "Uniform Location of Subject Matter," Document Number 1910-16.
2. Construction Specifications Institute:
   a. *Manual of Practice* (2 vols.)
   b. Specification Document Series, Divisions 1 through 16.
3. U.S. Treasury Department, Audit Staff Bureau of Accounts Circular 570, "Companies Holding Certificates of Authority as Acceptable Sureties on Federal Bonds and as Acceptable Reinsuring Companies."
4. Clark, J.R., "Commentary on Contract Documents," NSPE Document No. 1910-9, Professional Engineers in Private Practice, National Society of Professional Engineers, 1974.
5. Hecker, M.J., "Specifying Pipeline Construction", *ASCE Transportation Engineering Journal*, TE4, July 1979, pp. 361-371.

CHAPTER 11

CONSTRUCTION METHODS

A. INTRODUCTION

This chapter discusses construction methods in common use. Local conditions, of course, will dictate variations, and the ingenuities of the owner, engineer, and contractor must be applied continually if construction costs are to be minimized and acceptable standards are to be achieved.

Commencement of the construction phase normally introduces the third party, the contractor, to the sanitary sewer project, and the division of responsibility and liability must be understood by all. The role of the engineer changes from active direction and performance during design (a relationship to the owner), to that of professional and technical observation during construction. His duties during construction (during the contractual relationship between owner and contractor) will determine for the owner that the work is substantially in accordance with the contract documents — in other words, acceptable. It is important to note, however, that the engineer's representative on the site is not expected to duplicate the detailed inspection of material and workmanship properly delegated to the manufacturer, supplier, and contractor.

Preconstruction conferences are helpful in deciding whether the contractor's proposed operations are compatible with contract requirements and whether they will result in finished construction acceptable to the owner and the engineer. These joint meetings of the owner, engineer, and contractor should culminate in definite construction plans and administrative procedures to be followed throughout the life of the construction contract in accordance with that contract and should include items such as progress schedules, payment details, method of making submittals for approval, and channels of communication. All of these aspects of construction should be mutually understood before construction begins.

B. CONSTRUCTION SURVEYS

1. General

Baselines and benchmarks for sanitary sewer line and grade control should be established along the route of the proposed construction by the engineer. All control points should be referenced adequately to permanent objects located outside normal construction limits.

2. Preliminary Layouts

Prior to the start of any work, rights-of-way, work areas, clearing limits, and pavement cuts should be laid out to give proper recognition to, and protection for, adjacent properties. Access roads, detours, bypasses, and protective fences or barricades also should be laid out and constructed as required in advance of sanitary sewer construction. All layout work, if done by the contractor, should be reviewed by the engineer before any demolition or construction begins.

3. Setting Line and Grade

The transfer of line and grade from control points established by the engineer to the construction work should be the responsibility of the contractor, with spot checks by the engineer as work progresses. The preservation of stakes or other line and grade references provided by the engineer is similarly the responsibility of the contractor. (Generally a charge is made for re-establishing stakes carelessly destroyed, and the charge is stated as part of the contract agreement.)

In general, the line and grade for the sanitary sewer may be set by one or a combination of the following methods:

(a) Stakes, spikes, or crosses set at the surface on an offset from the sanitary sewer centerline.

(b) Stakes set in the trench bottom on the sanitary sewer line as the rough grade for the sanitary sewer is completed.

(c) Elevations given for the finished trench grade and sanitary sewer invert while sanitary sewer laying progresses.

(d) Laser beam of light set in the manhole.

The first method generally is used for small diameter sanitary sewers. Methods (b) and (c) are used for large sanitary sewers or where sloped trench walls result in top-of-trench widths too great for practical use of short offsets or batter boards. Method (d) is independent of the size of sanitary sewer.

In the first method, stakes, spikes or crosses are set on the opposite side of the trench from which excavated materials are to be cast at a uniform offset, insofar as practicable, from the sanitary sewer centerline. A cut sheet is prepared (Table 11-1) which is a tabulation of the reference points giving sanitary sewer station, offset, and the vertical distance from reference point to proposed sanitary sewer invert.

The line and grade may be transferred to the bottom of the sanitary sewer trench by the use of batter boards, tape and level, or patented bar tape and plumb bob unit.

Batter boards and batter-board supports must be suspended firmly across the trench and be adequate to span the excavation without measurable deflection. If the spanning member is to be the batter board, it is set level at an even foot (or other unit of measurement) above sanitary sewer grade and a nail is driven in the upper edge at centerline of the sanitary sewer. Preferably, the spanning member is used as a support only and a 2.5-cm (1-in.) board is nailed to it with one edge in a true vertical plane at the centerline of the sanitary sewer. A nail then is driven in the vertical edge of the batter board at an even foot above sanitary sewer grade. A string line is drawn taut across at least three batter boards.

The sanitary sewer centerline is then transferred to the trench bottom with a heavy plumb bob cord held lightly against the string line. Grade is transferred to the sanitary sewer invert with a grade rod equipped with a suitable metal foot to extend into the end of the pipe. For steep grades it is advisable to fasten a bullseye level to the grade rod to assure that the rod is held plumb. For ease in reading, the grade rod may be marked at subgrade, finish grade, and invert grade with saw grooves, small nails, or colored bands. If nails or grooves are used, care must be taken in reading the rod to insure that the string line is not riding up or down on the reference devices. The line and grade of the string line should be checked by observation for possible error in cuts or in establishing the batter boards. Periodic inspection should be made during sanitary sewer laying to insure that the set line and grade have not been disturbed.

Table 11-1 Typical Cut Sheet

"A" street .... sewer Sheet 1 of 1 Sheet
4th Street to 5th Street   Notes book .... Page .....
Stakes 5 ft left   Prepared by .... Date .....

Station (1)	Size, in inches (2)	Grade (3)	Elevation		Cut		Remarks (8)
			Invert (4)	Stake (5)	Feet (6)	Hundredths (7)	
0+00	12	0.0025	105.50	110.75	5	25	Existing manhole
0+25	..	0.0025	105.56	112.30	6	74	..
0+50	..	0.0025	105.63	109.70	4	07	Y-branch right
0+75	..	0.0025	105.69	110.35	4	66	..
1+00	12	0.0025	105.75	111.99	6	24	Etc.

Note: Ft × 0.3 = m; in. × 2.54 × cm.

Another method of setting grade is from offset crosses or stakes or from offset batter boards and double string lines and the use of a grade rod with a target near the top. When the sanitary sewer invert is on grade, a sighting between grade rod and two or more consecutive offset bars or the double string line will show correct alignment.

The transfer of surface references to stakes along the trench bottom is in some instances permitted, but the use of batter boards is preferred. If stakes are established along the trench bottom, a string line should be drawn between not less than three points and checked in the manner used for batter boards.

When trench walls are not sheeted but sloped to prevent caving, line-and-grade stakes are set in the trench bottom as the excavation proceeds. This procedure requires a field party to be at the work almost constantly.

A third method, applicable to large diameter sanitary sewers or monolithic sections of sanitary sewers on flat grades, requires the line and grade for each pipe length or form sections to be set by means of a transit and level from either on top or inside of the completed conduit.

In the construction of large sanitary sewer sections in an open trench, both line and grade may be set at or near the trench bottom. Line points and benchmarks may be established on cross bracing where such bracing is in place and rigidly set. Later, alignment and grade must be determined by checking the setting of the forms.

A fourth method, which is quite widely used is the laser beam control. A laser is a device which projects a narrow beam of light down the centerline of the sewer pipe. It is usually set up in the invert of a manhole and then aligned horizontally. The proper slope is established by adjusting a dial on the machine and aiming the laser. A check elevation should be set about 30 m (100 ft) from the manhole to assure that the proper slope is being maintained by the beam of light. A target set in the pipe centerline is then used to align the end of each pipe section. Care should be exercised in the use of the laser since temperature affects the aiming of the unit.

4. Tunnel Construction

Where tunnel construction is an extension of a sanitary sewer of sufficient size without change of alignment, the initial line and grade for the tunnel work may be established by extending lines and grades through and forward from the completed portion.

When tunneling begins from an isolated shaft, great care must be taken in transferring line and grade from the surface. If tunneling from any one shaft extends more than several hundred feet (meters) from the shaft, and especially if the alignment is curvilinear, it may be desirable to drive vertical line pipes to the elevation of the tunnel at intervals of 183 m (600 ft) or less through which plumb lines can be dropped when excavation reaches those points. This will allow a check of both alignment and grade at each line pipe location. If tunneling is to be done under air pressure, these pipes must be capped at either the top or bottom at all times. All line pipes should be grout-sealed on completion of construction.

As cleanup progresses in preparation for final lining operations, cross-sections should be taken throughout the tunnel to insure that proper alignment and grade can be maintained. Tunnel geology and cross-sections should be recorded for incorporation in permanent construction records.

C. SITE PREPARATION

The amount of site preparation required varies from none to the extreme where the major portion of project costs is expended on items other than excavating for and construction of the sanitary sewer.

Operations which may properly be classified as site preparation are clearing and grubbing removal of unsuitable soils; construction of access roads, detours, and bypasses; improvements to and modification of existing drainage; location, protection, or relocation of existing utilities and pavement cutting. The extent and diversity of these operations make further discussion thereof impractical. Note, however, that the success of the contractor in keeping the project on schedule depends to a great degree on the thoroughness of the planning and execution of the site preparation work. Several engineers and contractors have adopted a practice of assembling extensive photographic or video-tape evidence of pre-construction condition of sidewalks, driveways, street surfaces, etc., to minimize post-construction claims by residents for construction related damages.

D. OPEN-TRENCH CONSTRUCTION

1. Trench Dimensions

With plans and specifications competently prepared it can be assumed that the location of the proposed sanitary sewers has been determined with proper regard for the known locations of existing underground utilities, surface improvements, and adjacent buildings. Barring unforeseen conditions, it becomes the objective of the contractor to complete the work as shown on the plans at minimum cost and with minimum disturbance of adjacent facilities.

Because of load considerations as discussed in Chapter 9, the width of trench at and below the top of the sanitary sewer should be only as wide as necessary for proper installation and backfilling, and consistent with safety. The contract must provide for alternate methods or require corrective measures to be employed by the contractor if allowable trench widths are exceeded through overshooting of rock, caving of earth trenches, or over-excavation.

The width of trench from a plane 30 cm (1 ft) above the top of the sanitary sewer to the ground surface is related primarily to its effect on the safety of the workmen who must enter the trench and on adjoining facilities, such as other utilities, surface improvements, and nearby structures.

In undeveloped subdivisions and in open country, economic considerations often justify sloping the sides of the trench for earth stability from a plane 30 cm (1 ft) above the top of the finished sanitary sewer to the ground surface. This eliminates placing, maintaining, and removing substantial amounts of temporary sheeting and bracing, unless safety regulations make some type of sheeting or bracing mandatory. Steel trench shields are often used to protect the workmen where sheeting is deemed unnecessary.

In improved streets, on the other hand, it may be desirable to restrict the trench width so as to protect existing facilities and reduce the cost of surface restoration. Available working space, traffic conditions, and economics will all influence this decision.

2. Excavation

With favorable ground conditions excavation can be accomplished in one simple operation; under more adverse conditions it may require several steps. In general, stripping, drilling, blasting, and trenching will cover all phases of the excavation operation.

In all excavations extreme care should be taken to properly locate, support and protect existing utilities. The owners of the utilities should be contacted before the start of excavation.

a. Stripping

Stripping may be advantageous or required as a first step in trench excavation for a variety of reasons, the most common of which are:

(1) Removal of topsoil or other materials to be saved and used for site restoration;

(2) Removal of material unsatisfactory for backfill to insure its separation from usable excavated soils;

(3) Removal of material having a low bearing value to a depth where there is material capable of supporting heavy construction equipment;

(4) Reduce cuts to depths down to which a backhoe can dig;

(5) Make it easier to charge drill holes.

b. Drilling and Blasting

In some areas sanitary sewer and house connections must be installed in hard rock. In addition, some shales and softer rocks which may be ripped in open excavation will require blasting before they can be removed in confined areas as required in trenching or excavating for structures.

Normally the most economic method will involve preshooting; that is, drilling and shooting rock before removal of overburden. In some instances the occurrence of wet granular materials above the rock ledge will necessitate stripping before drilling, since holes cannot be maintained open through the overburden to permit placing of explosive charges.

For narrow trenches in soft rock, a single row of drill holes may be sufficient. One or more additional rows will be required in harder rock and for wider trenches. To reduce overbreak and improve bottom fragmentation, time delays should be used in blasting for trenches. In tight quarters, trench walls can be presplit and the center shot in successive short rounds to an open face to produce minimum vibration.

It must be recognized that there will be a minimum feasible trench width varying with the rock formation, and in the case of small sanitary sewers it may be necessary to design the conduit for the positive projecting condition or extreme loads.

All ground and air pressures that result from blasting should be recorded on a sealed cassette seismograph. Surveys of adjacent structures for the instance of cracks before the blasting is started should be considered. Blasting should be done only by persons experienced in such operations.

c. Trenching

The method and equipment used for excavating the trench will depend on the type of material to be removed, the depth, amount of space available for operation of equipment and storage of excavated material, and prevailing practice in the area.

Ordinarily the choice of method and equipment rests with the contractor. However, various types of equipment have practical and real limitations regarding minimum trench widths and depths. The contractor is obligated, therefore, to utilize only that equipment capable of meeting trench width limitations imposed by sewer pipe strength requirements or for other reasons as set forth in the technical specifications.

Spoil should be placed sufficiently back from the edge of the excavation to prevent caving of the trench wall and to permit safe access along the trench. With sheeted trenches, a minimum distance of 1 m (3 ft) from the edge of sheeting to the toe of spoil bank will normally provide safe and adequate access. Under such conditions the supports must be designed for the added surcharge. In unsupported trenches the minimum distance from the vertical projection of the trench wall to the toe of the spoil bank normally should be not less than one-half the total depth of excavation. In most soils, this distance will be greater in order to provide safe access beyond the sloped trench walls.

Trenching Machines

This type of machine is generally used for shallow trenches less than 1.5 m (5 ft) deep. For installation of small sanitary sewers and force mains in cohesive soils, the trenching machine can make rapid progress at low cost.

Backhoes

These machines (Fig. 11-1) are available with bucket capacities varying from 0.3 to 2.3 m^3 (3/8 to 3 cu yd) and more. They are convenient for the excavation of trenches with widths exceeding 0.7 m (2 ft) and to depths down to 8 m (25 ft); they are the most satisfactory equipment for excavation in loosened rock. Minimum trench widths are compared with some common backhoe sizes in Table 11-2.

The backhoe also is used with a cable sling for lowering sewer pipe into the trench. By this means a single piece of equipment can maintain the sewer pipe laying close to the point of excavation. Where the soil does not require sheeting and bracing, this method becomes a very economical one. When sheeting and bracing must follow the excavation closely, the combination of a backhoe for excavation and a crane for placement of sewer pipe is a common practice.

Clamshells

When the protection of other underground structures or soil conditions require close sheeting and the use of vertical-lift equipment, the clamshell bucket is used. In very deep trenches where two-stage excavation is required,

Fig. 11-1. Typical backhoe.

the backhoe is sometimes used in combination with the clamshell, with the backhoe advancing the upper part of the excavation and the clamshell following for the lower. Sheeting and bracing of the upper part are installed and driven as required prior to the clamming of the lower part and the installation of the lower-stage of sheeting.

Draglines

In open country for stream crossings or in a wide right-of-way, it may be feasible to do a large part of the excavation by means of a dragline, allowing

Table 11-2 Trench Widths Associated with Various Backhoe Bucket Capacities

Bucket Capacity, in cubic yards (1)	Minimum Trench Width, in inches	
	Without Side Cutters (2)	With Side Cutters (3)
⅜	22	24–28
½	27	28–32
¾	28	28–38
1	34	34–44
1¼	37	37–46
1½	38	38–46
2	50	50–58

Note: in. × 2.54 = cm; cu yd × 0.76 = m^3

the sides of the trenches to acquire their natural slope. In cases of very deep trench excavation, say 9 to 15 m (30 to 50 ft), the dragline has been used for the upper part of the excavation, with a backhoe operating at an intermediate level. By rotating the backhoe, the material thus excavated can be relayed to the dragline, which then lifts it to the spoil bank or to trucks at the surface.

Front-End Loaders

In wide, deep trenches the front-end loader has sometimes been used as an auxiliary to a backhoe or clamshell. In this arrangement the backhoe excavates the upper part of the trench and, perhaps, the center section of the lower part, leaving the bottom bench or benches for the front-end loader or dozer which completes the excavation, placing the spoil within the reach of the backhoe or clamshell.

The principal use of loaders, however, is in bringing sewer pipe, manholes and granular bedding material to the trenches.

3. Sheeting and Bracing

Trench sheeting and bracing should be adequate to prevent cave-in of the trench walls or subsidence of areas adjacent to the trench and to prevent sloughing of the base of the excavation from water seepage. Responsibility for the adequacy of any required sheeting and bracing usually is stipulated to be with the contractor. The strength design of the system of supports should be based on the principles of soil mechanics and structural engineering as they apply to the materials encountered. Sheeting and bracing always must comply with applicable safety requirements.

For wider and deeper trenches a system of wales and cross struts of heavy timber (or steel sections), as shown in Fig. 11-2, often is used. Sheeting is installed outside the horizontal wales as required to maintain the stability of the trench walls. Jacks mounted on one end of the cross struts maintain pressure against the wales and sheeting.

In some soil conditions it has been found economical and practical to use steel trench shields which are pulled forward as sewer pipe laying progresses. Care must be exercised in pulling shields forward so as not to drag or otherwise disturb the previously laid pipe sections or to create conditions not assumed in calculating trench loads.

In non-cohesive soils containing considerable groundwater, it may be necessary to use continuous steel sheet piling to prevent excessive soil movements. Such steel piling usually extends several feet (meters) below the bottom of the trench unless the lower part of the trench is in firm material. In the latter case the width of the trench in the upper granular material may be widened so that the steel sheet piling can toe-in to the lower strata.

In some soils, steel sheet piling (Fig. 11-3) can be used with a backhoe operation for the upper part of excavation, but the piling usually needs to be braced before the excavation has reached its full depth, and the remaining excavation then is performed by vertical life equipment such as a clamshell.

Another means of trench sheeting occasionally adopted involves the use of vertical H-beams as "soldier beams" with horizontal wood lagging. This is sometimes advantageous for trenches under existing overhead viaducts where overhead clearances are low and spread footings lie alongside the trench walls. The holes for the vertical beam can be tilted into these holes and driven. As excavation progresses downward, the lagging is installed between adjacent pairs of soldier beams. For deep trenches with limited overhead clearances, the soldier beams can be delivered to the site in shorter lengths and their ends field-welded as driving progresses.

Fig. 11-2. Sheeting and bracing system.

Fig. 11-3. Steel sheeting in approximately 20-ft (6-m) deep trench.

The removal of sheeting following pipe laying may affect the earth load on the sanitary sewer or adjacent structures. This possibility must be considered during the design phase. If removal is to be permitted, appropriate requirements must be placed in the technical specifications. Whenever sheeting is removed it must be done properly, taking care to backfill thoroughly the voids thus created.

4. Dewatering

Trenches should be dewatered for concrete placement and sewer pipe laying, and they should be kept continuously dewatered for as long as necessary. Unfortunately, the disposal of large quantities of water from this operation, in the absence of storm drains or adjacent water courses, may present problems. Other means of disposal being unavailable, the possibility of draining the water through the complete sanitary sewer to a permissible point of discharge should be considered, provided sufficient precautions are taken to prevent scour of freshly placed concrete or mortar or transport of sediment. In all cases care must be exercised to insure that property damage, including silt deposits in sanitary sewer and on streets, does not result from the disposal of trench drainage.

Crushed stone or gravel may be used as a subdrain to facilitate drainage to trench or sump pumps. It is good practice to provide clay dams in the subdrain to minimize the possibility of undercutting the sewer foundation from excessive groundwater flows.

An excessive quantity of water, particularly when it creates an unstable soil condition, may require the use of a well-point system. A system of this type consists of a series of perforated pipes driven or jetted into the water-bearing strata on either side of the sanitary sewer trench and connected with a header pipe leading to a pump. The equipment for a well-point system is expensive and specialized. General contractors often seek the help of special dewatering contractors for such work. Well-point systems must be run continuously to avoid disturbing the excavated trench bottom by uplift pressure.

Where excavation is in coarse water-bearing material, turbine well pumps may be used to lower the water table during construction. Chemical or cement grouting and freezing of the soil adjacent to the excavation have been used in extremely unstable water-bearing strata.

Water from all types of dewatering systems should be checked to assure that fine-grained material is not being removed from beneath the pipe, which might cause future settlement.

5. Foundations

Firm cohesive soils provide adequate sewer pipe foundations when properly prepared. Occasionally the trench bottom may be shaped to fit the sewer pipe barrel and holes dug to receive projecting joint elements. It is often a practice to over-excavate and backfill with granular material, such as crushed stone, crushed slag, or gravel, to provide uniform bedding of the sewer pipe. Such granular bedding is used because it is both practical and economical.

In very soft bottoms it is necessary first, as a minimum, to overexcavate to greater depths and stabilize the trench bottom by the addition of gravel or crushed slag or rock compacted to receive the load. The stabilizing material must be graded to prevent movement of subgrade up into stabilized base and base into bedding material. There is increasing use of specialized filter fabrics to prevent this movement. The required depth of stabilization should be determined by tests and observations on the job.

In those instances where the trench bottom cannot be stabilized satisfactorily with a crushed rock or gravel bed, and where limited and intermittent areas of unequal settlement are anticipated, a timber cribbing, piling, or reinforced concrete cradle may be necessary.

Where the bottom of the trench is rock, it must be overexcavated to make room for an adequate bedding of granular material which will uniformly support the conduit. The trench bottom must be cleaned of shattered and decomposed rock or shale prior to placement of bedding.

In some instances, sanitary sewer pipe (i.e. not cast-in-place) must be constructed for considerable distances in areas generally subject to subsidence, and consideration should be given to constructing them on a timber platform or reinforced concrete cradle supported by piling. The sewer's support should be adequate to sustain the weight of the full sewer and backfill. Piling in this case is sometimes driven to grade with a follower prior to making the excavation. This practice avoids subsidence of trench walls resulting from pile driving vibrations. Extreme care must be taken to locate all underground structures.

6. Pipe Sanitary Sewers

Proper sanitary sewer construction implies that quality materials and acceptable laying methods are to be used. Diligence in assuring both is required of all project personnel.

a. Sewer Pipe Quality

Sewer pipe inspection is properly conducted by the manufacturer and by independent testing and inspection laboratories. Moreover, with pipe sanitary sewers, transportation charges may constitute a substantial portion of material costs and inspection at the pipe plant, therefore, is usually desirable. Inspection may consist of visual inspection of workmanship, surface finish, and markings; physical check of length, thickness, diameter, and joint tolerances; proof of crushing strength (rigid pipe) or pipe stiffness (flexible pipe) design material tests, and tests of representative specimens. If three-edge bearing tests are not used on precast concrete pipe, core or cylinder tests should be required. However standard cylinder tests are not practical with the mixes used in some manufacturing methods and core tests are generally used. Cores also permit checking tolerances on placement of reinforcing cages.

Sewer pipe suppliers should furnish certificates of compliance with specifications that can be easily checked as the loads of sewer pipe arrive at the site.

Sewer pipe also should be checked visually at time of delivery for possible damage in transit and again as it is laid for damage in storage or handling.

b. Sewer Pipe Handling

Care must be exercised in handling and bedding all precast sewer pipe, regardless of cross-sectional shape. All phases of construction should be undertaken to insure that, insofar as practical, pipe is installed as designed. Sewer pipe should be handled during delivery in a manner which eliminates any possibility of high impact or point loading, with care taken always to protect joint elements.

c. Sewer Pipe Placement

Sewer pipe should be laid on a firm but slightly yielding bedding true to line and grade, with uniform bearing under the full length of the barrel of the sewer pipe, without break from structure to structure, and with the socket ends of bell and spigot or tongue and groove sewer pipe joint facing upgrade. Sewer pipe should be supported free of the bedding during the jointing process to avoid disturbance of the subgrade. A suitable excavation should be made to receive sewer pipe bells and joint collars where applicable so that the bottom reaction and support are confined only to the sewer pipe barrel. Adjustments to line and grade should be made by scraping away or adding adquately compacted foundation material under the sewer pipe and not by using wedges and blocks or by beating on the sewer pipe.

Extreme care should be taken in jointing to insure that the bell and spigot are clean and free of any foreign materials. Joint materials vary with the type of sewer pipe used (see Chapter 8). All pipe joints should be made properly using the jointing materials and methods specified. All pipe joints should be sufficiently tight to meet infiltration or exfiltration tests.

In large diameter sewers with compression-type pipe joints, considerable force will be required to insert the spigot fully into the bell. Come-alongs and winches or the crane itself may be rigged to provide the necessary force (Fig. 11-4). Inserts should be used to prevent the sewer pipe from being thrust completely home prior to checking gasket location. After the gasket is checked, the inserts can be removed and the joint completed.

The operation of equipment over small diameter sewer pipe, or other actions that would otherwise disturb any conduit after pipe jointing must not be permitted.

At the close of each day's work, or when sewer pipe is not being laid, the end of the sewer pipe should be protected by a close-fitting stopper to keep the pipe clean, with adequate precautions taken to overcome possible uplift. The elevation of the last sewer pipe placed should be checked the next morning before work resumes.

If the sewer pipe load factor is increased with either arch or total encasement, contraction joints should be provided at regular intervals in the encasement coincident with the pipe joints to increase flexibility of the encased conduit.

7. Backfilling

a. General Considerations

Backfilling of the sanitary sewer trench is a very important consideration in sanitary sewer construction and seldom receives the attention and inspection it deserves. The methods and equipment used in placing fill must be selected to prevent dislocation or damage to the sewer pipe.

The method of backfilling varies with the width of the trench, the character of the materials excavated, the method of excavation, and the degree of compaction required.

b. Degree of Compaction

In improved streets or streets programmed for immediate paving, a high degree of compaction should be required. In less important streets or in sparsely inhabited subdivisions where flexible macadam roadways are used, a more moderate specification for backfilling may be justified. Along outfall sewers in open country, it may be sufficient to mound the trench and, after natural settlement, return to regrade the area. Compaction results should be determined in accordance with current AASHTO or ASTM test procedures. Laboratory tests to establish optimum moisture content are commonly done

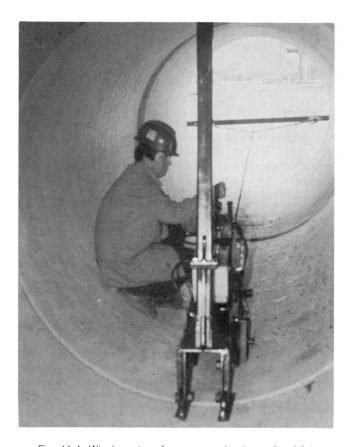

Fig. 11-4. Winch system for compression-type pipe joint.

according to modified Proctor method, AASHTO T180 or ASTM D1557. Field tests to determine actual compaction may be done by any of several mechanical methods, or by the use of nuclear density meters.

c. Methods of Compaction

Cohesive Materials

Cohesive materials with high clay content are characterized by small particle size and low internal friction. They have small ranges of moisture content over which they may be compacted satisfactorily and are very impervious in a dense state. Because of the strong adhesive forces of the soil particles, strong pressures must be exerted in order to shear the adhesive forces and remold the particles in a dense soil mass. These characteristics dictate the use of impact-type equipment for most satisfactory results in compaction. In confined areas, pneumatic tampers and engine-driven rammers may give good results. The upper portion of the trench can be consolidated by self-propelled rammers where trench widths are relatively narrow. In wide trenches sheepsfoot rollers may be used; if the degree of compaction required is not high, dozers and loaders may be used to compact the fill.

Regardless of equipment used, the soil must be near optimum moisture content and compacted in multiple lifts if satisfactory results are to be obtained. The trench bottom must be free of excessive water before the first lift of backfill is placed.

If the material had a high moisture content at the time of excavation, some preparation of the material probably will be required before spreading in the trench. This may include pulverizing, drying, or blending with dry or granular materials to improve placement and consolidation.

Cohesionless Materials

Cohesionless materials are granular with little adhesion but with high internal friction. Moisture content at the time of compaction is not so critical, and consolidation is affected by reducing the surface friction between particles thus allowing them to rearrange in a more compact mass. Considering the characteristics of cohesionless material, the most satisfactory compaction is achieved through the use of vibratory equipment.

In confined areas, vibratory plates give the best results. For wider trenches vibratory rollers are most satisfactory. Again, if the degree of compaction required is not high, and if layers are thin, the vibration imparted by dozer or loader tracks may result in satisfactory consolidation.

In some areas water is used to consolidate granular materials. But unless the fill is saturated and immersion vibrators are used, the degree and uniformity of compaction cannot be controlled closely. With some materials, adequate compaction may be obtained by draining off, through drains constructed in manhole walls, water used to saturate or puddle fill. These drains are capped after the backfill has drained.

Borrow Materials

Sometimes the material removed from the trench may be entirely unsatisfactory for backfill. In this case, selected materials must be hauled in from other sources.

Both cohesive and noncohesive materials are used, but an assessment must be made of the possible change in groundwater movement that the use of outside materials may cause. For example, the use of cohesive materials to backfill a trench in rock could result in a dam inpervious to groundwater traveling in rock faults, seams, and crevices. On the other hand, granular materials placed in a clay trench could result in a very effective subdrain.

d. Backfilling Sequence

Backfilling should proceed immediately on curing of trench-made joints and after the concrete cradle, arch, or other structures gain sufficient strength to withstand loads without damage.

Backfill generally is specified as consisting of three zones with different criteria for each: The first zone (embedment) extends from the foundation material to 30 cm (12 in.) above top of sewer pipe or structure; an intermediate zone generally contains the major volume of the fill; and the upper zone consists of pavement subgrade or finish grading materials.

The first zone should consist of selected materials placed by hand or by suitable equipment in such manner as not to disturb the sewer pipe, and compacted to a density consistent with design assumptions. In some instances the material used for granular bedding is brought above the sewer pipe to insure high density backfill with minimum compactive effort. When installing flexible pipes, attention must be given to proper placement and compaction of the haunching material from the base of the pipe to the springline. When high water tables are anticipated, embedment materials without substantial voids are required to prevent soil migration. In Section D of Chapter 9, the effect of compaction in this first zone on the sewer pipe-soil system is discussed in greater detail.

Compaction of the intermediate zone is usually controlled by the location. Under traffic areas or other improved existing surfaces, 95% of modified Proctor density should be required. In other general urban areas, 90% may be adequate. In undeveloped areas, little compaction may be required. In general, the degree of compaction required will often affect the choice of material. The use of excavated material, if suitable, is usually best in areas which are subject to frost heave so that excavated areas will not move more or move less than undisturbed areas.

Depth and compaction of the upper zone are dependent on the type of finish surface to be provided. If the construction area is to be seeded or sodded, the upper 450 mm (18 in.) may consist of 350 mm (14 in.) of select material slightly mounded over the trench and lightly rolled, covered by 100 mm (4 in.) of top soil. If the area is to be paved, the upper zone must be constructed to the proper elevation for receiving base and paving courses under conditions matching design assumptions for the subgrade. If the trench backfill is completed in advance of paving, the top 150 mm (6 in.) of the upper zone should be scarified and recompacted prior to paving. In such instance, it may be necessary to install a temporary surface to be replaced at a later date with permanent pavement.

Before and during the backfilling of a trench, precautions should be taken against the flotation of pipelines due to the entry of large quantities of water into the trench, which could cause uplift on the empty or partly filled pipeline.

8. Surface Restoration

On completion of backfill the surface should be restored fully to a condition at least equal to that which existed prior to the sanitary sewer construction.

Portland cement or asphaltic concrete pavements should be saw cut and removed to a point beyond any caving or disturbance of the base materials prior to patching. If this results in narrow, unstable panels, pavement should be removed to the next existing contraction or construction joint. Before replacing permanent pavement, the subgrade must be restored and compacted until smooth and unyielding.

The final grade in unpaved areas should match existing grades at construction limits without producing drainage problems. Restoration of grass, shrubs, and other planting should be done in conformance with construction contract documents. Tree damage should be repaired in accordance with good horticultural practice.

E. TUNNELING

1. General Classification

Tunneling is considered to be any construction method which results in the placement or construction of an underground conduit without continuous disturbance of the ground surface, and includes the various forms of jacking of prefabricated units from shaft or pit locations. Tunneling methods applicable to sanitary sewer construction can be classified generally as follows:
(a) Auger or boring method;

(b) Jacking of preformed steel or concrete pipe; and
(c) Mining methods.

2. Auger or Boring

In sizes less than 900-mm (36-in.) diameter, rigid steel or concrete pipe can be pushed for reasonable distances through the ground and the earth removed by mechanical means under the control of an operator at the shaft or pit location. Several types of earth augers are available, and some contractors specialize in this type of operation. Augers as large as 1,800 mm (72 in.) have been used, but for sizes above 900 mm (36 in.) considerable care must be exercised to avoid overbreak. In the case of steel pipe, the auger need not be larger than the inside diameter. In the case of concrete pipe, it may be necessary to use an auger with a special head having a diameter equal to the outside diameter of the pipe being placed.

The presence of any concretions is a serious deterrent to this method of installation. If such obstructions are expected, particularly when sewer pipes smaller than 900 mm (36 in.) are to be placed, it may be more economical first to install an oversize lining by conventional tunnel or jacking methods. The sewer pipe then can be placed within the liner pipe and the remaining space backfilled with sand, cement grout, or concrete.

3. Jacking

Although the limits will vary with geographic locations and soil conditions, finished interior diameters of 750 to 2,750 mm (30 to 108 in.) are the generally accepted limits for pipe jacking. Excavation and removal of the excavated material is by machine or manual labor augmented with air spades, special knives, etc. The most commonly used materials for such jacking operations are reinforced concrete or smooth steel pipe. The pipe sewer selected for jacking must be strong enough to withstand the loads exerted by the jacking procedure.

The usual procedure is to equip the leading edge with a cutter or shoe to protect the sewer pipe. As succeeding lengths of pipe are added between the leading sewer pipe and the jacks and the sewer pipe jacked forward, soil is excavated by hand and removed through the sewer pipe. Material is trimmed with care and excavation does not precede the jacking operations more than necessary. Such a method usually results in minimum disturbance of the natural soils adjacent to the sewer pipe.

When jacking, contractors have sometimes found it desirable to coat the outside of the pipe with a lubricant, such as bentonite, to reduce the frictional resistance. In some instances this lubricant has been applied through pressure fittings installed in the wall of the leading pipe. Grout holes sometimes are provided in the walls of the pipes for use in filling outside voids. Protective joint spacers are used to prevent damage to pipe joints.

Because soil friction may increase with time, it is desirable to continue jacking operations without interruption until completed.

a. Alignment

In all jacking operations it is important that the direction of jacking be carefully established prior to the start of work and checked periodically during the work. Guide rails must be installed in the bottom of the jacking pit or shaft. In the case of a large pipe it is desirable to have such rails carefully set in a concrete slab. The number and capacity of the jacks used depend primarily

Fig. 11-5. Typical installation for jacking concrete pipe.

on the size and length of the pipe to be placed and the type of soil encountered.

Backstops must be strong enough and large enough to distribute the maximum capacity of the jacks against the soil behind them. A typical installation for jacking concrete pipe is shown in Fig. 11-5.

b. Continuous Sewers by Jacking

In some cases long sanitary sewer lines have been installed by jacking from a series of shaft locations spaced along the line of the sewer pipe.

4. Mining Methods

Tunnels with finished interior dimensions of 1.5 m (5 ft) or larger in clay or granular materials ordinarily are built either with the use of tunnel shields or with boring machines, or by open-face mining with or without some breasting. Rock tunnels normally are excavated open-face by conventional mining methods or with boring machines.

a. Tunnel Shields

In very soft clay or in running sand, especially in built-up city areas, it may be necessary to use tunnel shields to make the tunneling operation safe. Compressed air also may be required to control the entry of water into the tunnel.

With a shield it is necessary to install a primary lining of sufficient strength to support the surrounding earth and to provide a progressive backstop for the jacks which advance the shield. The lining may be installed against the earth and the annular space between the lining and the earth filled with pea gravel and grout. Alternatively, the tunnel lining may be expanded against the earth as the shield is advanced. The latter method practically eliminates the need for grouting the annular opening.

Tunnel shields may be of the open-face type, one that provides a hood under which miners excavate the tunnel heading. Other shield designs have ports in the face through which the material being removed flows. Where the displaced material is allowed to flow through ports in the face of the shield, a continuous check must be maintained on the volume of materials removed against the linear displacement. Street surface elevations must be observed continually for vertical movement either up or down.

Shield alignment usually is maintained by varying the thrust of the jacks around the periphery of the shield.

b. Boring Machines

Boring machines, also called digger shields or mechanical moles, have been developed for tunnel excavation in clay and rock. They usually have cutters mounted on a rotating head which is advanced into the heading. A conveyor system moves muck to the muck cars. Machines may be braced against the walls of the excavation or against previously placed tunnel lining. Some machines also are equipped with shields. Machines have been used successfully in the construction of tunnels in clay up to 6-m (20-ft) diameter.

Machines are most useful in fairly long runs through generally similar material. Difficulties have been encountered where the material to be excavated varies.

Shields are also being used economically for soft-ground tunnels in wet conditions, with a permanent, primary expanded lining of cast iron segments. The cast iron is covered with a concrete lining in free air, thus economizing on the more expensive compressed air labor. Such designs minimize street and adjacent building settlement as compared with temporary linings of steel liner plate.

c. Open-Face Mining Without Shields

Where the ground allows the use of open-face mining methods (Figs. 11-6 and 11-7) it is often more economical to use segmental supports of wood or steel for the sides and top of the tunnel only. The need for compressed air or breast boards in the tunnel heading will depend on the type of soil and amount of moisture or groundwater. Tunneling in soft clay in the presence of considerable groundwater usually would dictate the use of compressed air. Stiff clays which require the use of air spades for removal can be excavated normally by open-face methods without the use of compressed air.

d. Primary or Temporary Lining

Materials used for primary lining are usually steel, wood or a combination of the two. They also may be segmental blocks of concrete or cast iron.

Timber and Timber-and-Steel Combination

Tunnels have been timber supported with a continuous series of five-piece timber frames or cants kept close to the face of the tunnel. A combination of timber and steel often is used, in which case rib sets fabricated from I-beams, H-beams, or channels support wood lagging (Fig. 11-11). Rib sets may be fabricated full circle or have vertical legs with arch sections bent to the outside shape of the sewer. Ribs are set at required intervals, with intermediate support supplied by wood lagging placed outside the ribs or just inside the outer flanges. Encroachment of ribs within the neat line of the tunnel lining

Fig. 11-6. Open-faced mining methods.

Fig. 11-7. Open-faced mining methods.

section should not interfere with the placement of reinforcing steel, and generally is limited to 50 to 75 mm (2 to 3 in).

Steel

Many tunnels use only steel for the primary lining. Such steel lining has been in several forms and combinations. In some instances, liner plates bent to conform with the outside shape of the sanitary sewer section are bolted together along longitudinal and circumferential seams. Curved I-beam ribs, conforming to the outside of the sanitary sewer section, may be placed at transverse joints between any two successive rings of liner plates. In other cases, I-beam ribs are placed inside the liner plates, and the spacing of the ribs is varied according to pressure exerted by the soil on liner-plate sections. Deformed steel pans, or lagging plates, with flanges along two edges only, may be used to provide intermediate support between rib sets similar to that of wood lagging described previously. The plates are 300mm (12 in.) wide and vary in length from 600 to 1,500 mm (24 to 60 in.). Steel ribs are allowed to penetrate into neat lines of the concrete section 50 to 75 mm (2 to 3 in.). Concrete required to fill the space inside the flanges of the lagging plates is not considered a part of the finished sanitary sewer lining.

Some engineers and contractors prefer to use continuous rings of liner plates having sufficient section modulus to resist the earth pressures without use of special structure ribs or rings (Fig. 11-7). A circular lining formed of such plates becomes a compression ring and has some inherent stability not equaled by horseshoe-shaped supports. Soil conditions and the contractor's preference determine the choice of such a lining. Some liner plates have shallow corrugations with flanges on all four edges for bolting to the adjacent plates. Others have deeper corrugations with flanges only on the circumferential edges. Lapped, bolted joints are used along longitudinal edges of the latter type. Nuts can be welded to the exterior of the plate at bolthole locations, or the bolts can be held against rotation by special clips so that final tightening of the bolt can be accomplished from inside the tunnel. Steel liner plates frequently are used also as sheeting for circular shafts, either on or adjacent to the line of the tunnel.

If design of liner plate support is based on the assumption that plates will act as a compression ring, immediate grouting behind liner plates or immediate expansion of the lining is required to insure uniform loading. In any event, voids behind liner plates should be grouted or the lining expanded prior to subsidence of the overburden.

Precast Concrete

Precast concrete segments for primary lining units have been used in several different forms. Construction of a 3-m (10-ft) diameter sewer in Detroit was accomplished with seven-piece circular sets of precast blocks. Rings of primary lining were placed with the aid of a hydraulic placer arm mounted within the trailing part of a tunnel shield. Hydraulic jacks, mounted on the shield, are pushed against the completed rings of primary lining to advance the shield. A reinforced concrete lining was placed within the precast rings, resulting in a final section of approximately 300 mm (12 in.) of precast units and an additional 300 mm (12 in.) of concrete cast in place.

In a sanitary sewer of circular section in London, some precast concrete blocks formed with flanges and reinforcing ribs were used for primary lining. The finished lining was placed by shotcreting.

e. Oval Precast Concrete Rings

Another form of precast concrete tunnel lining consists of oval rings of short laying length. These sections are set with the long axis vertical. They are of such dimensions that they can be moved in horizontal position through previously placed sections, then rotated in the heading of the tunnel and jointed up with the last section placed. The oval rings are transported by a battery-operated locomotive specially designed for the purpose. Fig. 11-8, showing installation of oval precast concrete rings, does not indicate any primary lining. If soil conditions required initial primary lining, the lining should conform closely to the outside shape of the precast units. The space between precast units and primary lining must be filled with grout or other suitable material.

f. Tunnel Excavating Equipment

The type of excavating equipment or tools used in sanitary sewer tunneling is the same as in tunneling for other purposes and depends on the kind of material to be excavated and the work space available. Pneumatic spades and special knives are used widely in excavating clay. Drilling and blasting are necessary in rock tunnels. In the case of shale, undercutting machines like those used in coal mining have been used to advantage.

In a large sanitary sewer tunnel in Chicago, with a finished dimension of approximately 5.2 by 5.8 m (17 by 19 ft), an electric shovel with a 0.6-m^3 (3/4-cu yd) capacity dipper was used effectively. Excavated material usually is hauled from the face of the tunnel to the shaft location by means of small mucking cars pulled by electric locomotives.

g. Shafts

Where tunnels are of considerable length, one or more construction shafts need to be provided. On important thoroughfares these shafts are better located in an adjacent side street or vacant lot, with access to the work provided through a short connecting entry tunnel.

Offset shaft locations are especially desirable when soil conditions require the use of compressed air. In such a case only one air lock in the entry tunnel will be required. Shafts generally are located so that tunneling in both directions is possible. Construction shafts on long tunnels usually are spaced 350 to 750 m (1,200 to 2,500 ft) apart. Factors tending to extend this spacing are the need for compressed air, the size of tunnel, and the depth below ground.

Shafts should be large enough to permit the installation of an electric hoist. Such equipment should be used only for the handling of material, with separate facilities available for man-lifts in deep shafts. Hoisting in shafts by means of a crane may be permitted when the length of the tunnel is short and safety precautions are taken to prevent engine exhaust from entering the shaft and tunnel.

Tunnel drainage also is discharged normally from the shaft. Therefore, the shaft must contain some form of collection sump and drainage pump.

h. Main Shafts and Emergency Exits

Workmen should not use the material shaft for entry except when a part of it is provided with an enclosed stairway. Usually a separate shaft, adjacent to the material shaft, equipped with a spiral stairway is used for entry by workmen. Emergency hoisting facilities should be provided at shafts more than 15 m (50 ft) deep unless hoisting facilities in use are independent of power failures. Where feasible, emergency exits should be provided whenever

CONSTRUCTION METHODS

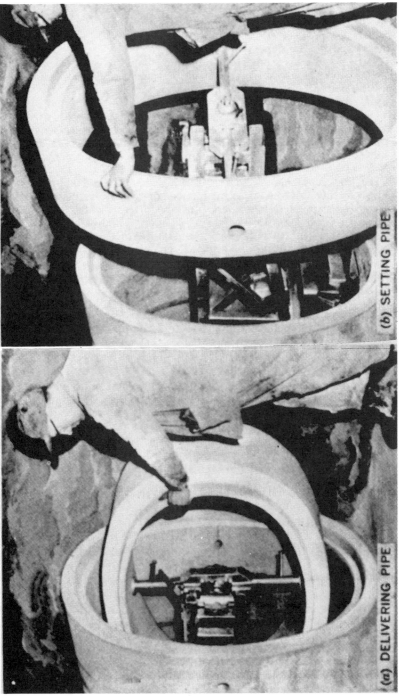

Fig. 11-8. Oval precast concrete rings.

the distance from a heading to the nearest exit exceeds 300 m (1,000 ft). If the tunnel is under air pressure, then each exit must be equipped with an air lock.

i. Compressed Air Equipment and Locks

Compressed air equipment for tunneling should have sufficient capacity to maintain a pressure which will prevent movement of the soil as well as to supply the normal air demand.

The equipment includes compressors, air receivers, piping, control valves, air locks, main and emergency locks, bulkhead walls, gages, etc. Separate locks should be provided for materials and personnel. Generally for long, large diameter tunnels, electrically operated compressors are used with two independent sources of power. Standby compressors in many cases are either diesel or gasoline powered.

The line supplying air to the working chamber is equipped with a pressure-reducing valve, a sufficient number of relief valves, and pressure-recording gages to insure a constant pressure in the chamber. Air locks are valved so that the lock may be operated either from within or without. Experienced lock tenders must be present at all times. Each lock door is equipped with a glass bullseye. Adequate signaling devices must be installed at each lock. A signal system for the compressor house, such as a siren operated from a mercoid switch on each of the pressure-reducing valves, should be installed. Air supplied to the working chamber should be of proper temperature and contain no objectionable substance.

If air pressure exceeds 100 kPa (15 psi) all necessary safeguards should be taken, including a medical lock and stage decompression facilities, with a physician on call day and night at all times that men are working.

j. Ventilating Air

In compressed air tunnels, air must be circulated in sufficient quantity to permit the work to be done without danger or excessive discomfort.

In free air tunnels, the ventilation rate must be adequate to clear the tunnel of gases in a maximum of 15 min if explosives are used. Rates also must be adequate to dilute exhaust of permissible diesel equipment to safe limits. A minimum of 5.5 m^3/min (200 cu ft/min) of fresh air per employee underground should be provided. In cold weather it may be necessary to condition ventilating air to prevent excessive fogging at the heading. Air should be monitored constantly for toxic or flammable gases and airborne contaminants. A record of all tests should be maintained.

F. SPECIAL CONSTRUCTION

1. Railroad Crossings

Sanitary sewers at times must be constructed under railroad tracks which may be at street level, on a raised embankment above street grade, or on an existing railroad viaduct.

Crossing of tracks at grade or on an embankment usually can be accomplished most economically by jacking, boring, tunneling, or a combination thereof. Usually a casing pipe is inserted and the sewer pipe is placed inside.

When the distance from the base of rail to the top of the sanitary sewer is insufficient to allow jacking or tunneling (less than one diameter clearance), it is necessary either to remove the tracks and interrupt service during an open-cut operation, or to build a temporary structure for support of the railroad tracks after which the sanitary sewer may be constructed in open trench below that structure.

Construction of sanitary sewers under existing railroad viaducts involves a wide variation in methods, depending on the size of the sanitary sewer, its location in plan and elevation with respect to viaduct footing, type of footings, and the nature of the soil.

Where the soil is stable and the sanitary sewer is of sufficient size to allow the use of tunnel methods and is located satisfactorily with respect to viaduct footings, tunneling may be both safe and economical.

When the proposed sanitary sewer does not meet these criteria, special methods of sheeting and bracing must be devised. To prevent subsequent movement of soil beneath the footings, all sheeting and bracing should be left in place.

Early planning with the railroad authorities is essential, since most companies have extensive design, inspection and permit requirements.

2. Crossing of Principal Traffic Arteries

Residential and secondary traffic arteries usually can be closed to traffic during the construction of sanitary sewer crossings. But on heavily traveled streets and highways where public convenience is a major factor it may be desirable to use tunneling or jacking methods for the crossing.

When required, limited traffic movements across open trenches can be accommodated by temporary decking. Trenches of narrow or medium width can be spanned with prefabricated decks placed on timber mudsills at the edges of the trench. Where the top of trench is wider than 5.3 to 6.7 m (16 to 20 ft) temporary piling for end support, and in some cases center support, may be required. Where center supports pass through the sanitary sewer section, provision must be made for such piling to remain until the sanitary sewer is completed. On a project in Chicago, center piling of steel was set on the centerline of a proposed twin-barrel sanitary sewer and later encased in the sanitary sewer section.

3. Stream and River Crossings

a. Sanitary Sewer Crossing Under Waterway

Stream and river crossings may be constructed either in the dry through use of cofferdams and diversion channels, or subaqueously. Open trenches may be excavated from barges with sewer pipe laying and pipe jointing done by divers. Ball joint sewer pipe may be effectively used, especially for force mains. For shallow stream crossings it may be possible to install an earth embankment and construct one-half the crossing at a time. If constructed in the dry, planning and scheduling of construction should be such that completed portions of the line are not subject to damage in the event of cofferdam overtopping.

Concrete encasement, if required, should be placed with construction joints at 9 to 12-m (30 to 40-ft) intervals coincident with pipe joints. Sewer pipe may be set conveniently to line and grade by supporting it on burlap bags filled with a dry-batched concrete mix. These bags also may be placed for construction of bulkheads in subaqueous concrete placements.

After placing the crossing beneath the bottom of the stream, it is usually advisable to place a layer of large rip-rap to form an armor course to protect the sewer pipe from erosion or hanging anchors.

Fig. 11-9. Installing prestressed concrete pipe beam section, Pittsfield, MA.

b. Sanitary Sewer Crossing Spanning Waterway

In Chapter 7 various methods of spanning obstacles are mentioned, including hanging and fastening sanitary sewers to structural supports and the construction of sanitary sewer pipe beams. The latter type of construction consists of a manhole or other supporting structure on either side of the waterway and the spanning member itself. Where the crossing is of considerable width, intermediate piers or supports may be necessary. Fig. 11-9 shows a 99-cm (39-in) diameter prestressed concrete pipe beam section 24 m (80 ft) long and of 22-m (72-ft) clear span, being lowered into final position over the west branch of the Housatonic River, Pittsfield, Mass.

4. Outfall Structures

a. Riverbank Structures

Sanitary outfall sewers and head walls may be located above or below surface water levels. When they are partly submerged it is necessary to provide some form of cofferdam during construction.

In shallow water, an earth dike or timber piling may be sufficient to maintain a dry pit. In deep water, steel sheet piling cofferdams are desirable. Usually a single wall cofferdam with adequate bracing is sufficient, but in excessive depths at the banks of main navigation channels, a double wall may be required. Standard practice of cofferdam design and construction should govern.

Fig. 11-10. Construction of 102-in. (260-cm) diam diffuser for ocean outfall. Note diffuser ports, a. Pipe is attached to "horse." Assemblage has swung clear of barge and is being lowered for placement at depth of 190 ft (58m). When "horse" has come to rest on bottom, man in control capsule, b, will use hydraulic controls to maneuver pipe into position.

b. Ocean Outfalls

For long ocean outfalls there are two distinct phases of construction: The inshore section through the surf zone and the offshore section. The surf zone usually extends to a depth of 15 m (50 ft) but may be shallower or deeper depending on local ocean conditions. The inshore or surf zone section requires positive support and lateral restraint for the outfall sewer pipe. The inshore section usually requires a temporary pier for driving sheet piling to maintain the trench through the breakers and for pipe installation. If the shore is all sand, suitable piles must be driven for support and anchorage of the outfall pipe.

The offshore section of outfall sewer pipe usually is laid from floating equipment (Fig. 11-10) and is often placed directly on the ocean floor provided the grade is satisfactory. Gravel or rock sidefill to the springline of the pipe is added frequently to prevent lateral currents from scouring local potholes which might cause pipe movement.

Ocean outfall pipes have been made of cast iron, reinforced concrete, protected steel pipe, plastic pipe and a combination of these materials. Small pipelines may be assembled on shore and then pulled or floated into position. Large lines must be laid in sections, although it is obviously advantageous to make the lengths as long as possible to minimize the number of underwater pipe joints which must be assembled by divers. It is also desirable to select a pipe joint type and construction procedure which will facilitate underwater connections. This is especially important if the outfall sewer is in water more than 50 m (150 ft) deep because of the limited time divers can work at this depth. Flexible pipe joint, ball-and-socket cast iron pipe laid with a cradle, permits the jointing of sewer pipe above water, thus eliminating the use of divers and underwater operations.

G. SEWER APPURTENANCES

Recent improvements in both sewer pipe and pipe joints have made it possible to construct a very watertight sanitary sewer system. It is important to be sure that large quantities of extraneous flows are not admitted at poorly constructed sanitary sewer appurtenances and service connections. Increased attention to this phase of sanitary sewer construction is essential if flow of surface and groundwaters in sanitary sewers is to be reduced.

The sound principles of construction which apply to reinforced concrete and masonry structures must be applied also to sanitary sewer manholes.

The use of watertight covers and proper waterproofing at the pipe joint between the frame and the top of the masonry also are very important.

Sanitary sewer connections should be permitted only at wye or tee branches or at machine cut, watertight-jointed taps. The fitting should be supported adequately during and after the pipe joint is made. Bell and spigot, compression-type flexible pipe joints should be used at the junction of the house sewer and service tap.

Caps and plugs for any deadend branches or house service connections should be made as watertight as any other pipe joint and be anchored to hold against internal pressure or external force.

H. CONSTRUCTION RECORDS

It is the responsibility of the engineer to record details of construction as accomplished in the field. These data should be incorporated into a final revision of the contract drawings to represent the most reliable record for future use.

Records should be sufficient to allow future recovery of the sewer itself, underground structures, connections, and services.

Invert elevations should be recorded for each manhole, structure, connection, and house service. In some instances, it may be advisable to set concrete reference markers flush with finished grade to facilitate future recovery.

I. REFERENCES

1. *Concrete Pipe Design Manual*, Amer. Concrete Pipe Assn., Vienna, VA.
2. *Design and Construction of Concrete Sewers*, Portland Cement Assn., Chicago, IL., 1968.
3. *Installation of Concrete Pipe*, Manual M9., Amer. Water Works Assn., Denver, CO 1979.
4. *Clay Pipe Engineering Manual*, National Clay Pipe Inst., Crystal Lake, IL.
5. "Recommended Practice for Installing Vitrified Clay Sewer Pipe," Std. C-12, Amer. Soc. Test. Mat., Philadelphia, PA.
6. "Installation of Gray and Ductile Cast Iron Water Mains and Appurtenances," Std. C-600, Amer. Water Works Assn., Denver, CO.
7. "Installation of Asbestos Cement Pressure Pipe," Std. C-603, Amer. Water Works Assn. Denver, CO.

8. "Code for the Manufacture, Transportation, Storage and Use of Explosives and Blasting Agents," Publ. No. 495, Natl. Fire Protection Assn.
9. *Blasters Handbook*, E. I. du Pont de Nemours and Company, Inc., Wilmington, DE, 1958.
10. *Manual of Accident Prevention in Construction*, Associated General Contractors, Washington, DC.
11. "Some Essential Safety Factors of Tunneling," Bur. Mines Bull. 439, U.S. Govt. Printing Office, Washington, DC.
12. "Mine Gases and Methods of Detecting Them," Bur. Mines, Miner's Circ. 33., U.S. Govt. Printing Office, Washington, DC.
13. "Protection Against Mine Gases," Bur. Mines, Miner's Circ. 35, U.S. Govt. Printing Office, Washington, DC.
14. "Engineering Factors in the Ventilation of Metal Mines," Bur. Mines Bull. 385, U.S. Govt. Printing Office, Washington, DC.
15. "Safety with Mobile Diesel Powered Equipment Underground," Bur. Mines Publ. R15616, U.S. Govt. Printing Office, Washington, DC.
16. *Handbook of PVC Pipe – Design and Construction*, Uni-Bell Plastic Pipe Association, Dallas, TX, 1981.

INDEX

Above ground sewers, 145-146
Acrylonitrile-Butadiene-Styrene (ABS) composite pipe, 159
Acrylonitrile-Butadiene-Styrene (ABS) pipe, 158-159
Air jumpers, 144-145
Air pressure testing, 124
Alternate depths, 86-88
Appurtenances, 108-109, 270
Arch analysis, 187-188
Asbestos Cement Pipe (ACP), 154
Auger, 259

Backfill, 216-218, 221
Backfilling, 193-196, 198-199, 255-258
Backflow preventors, 141
Backhoes, 249-250
Backwater gates, 145
Bedding, 193-195, 199-200, 201-204, 216-218, 221
Bends, 136-137
Bidding, 236-237
Bituminous pipe joints, 163
Blasting, excavation, 248-249
Boring, 259, 261
Borrow materials, 257
Building drain *(def.)*, 2
Building sewer *(def.)*, 2

Capacity estimates, 103-107
Cast Iron Pipe (CIP), 154-155
Cement mortar pipe joints, 163
Check valves, 141
Chezy equation, 93-94
Clamshell buckets, 249-250
Clean Water Act, 7
Coatings, 59-60
Cohesionless soils, 257
Cohesive soils, 256-257
Combined sewer *(def.)*, 1
Combined sewers, 113
Commercial areas, sewer needs, 31-33
Compaction, 255-257
Compressed air equipment, 266
Concentrated loads, 189-191
Concrete arches, 203-204
Concrete pipe, 64-66, 155-156
Conduit material, 101
Conduit shape, 101
Conduit size, 101
Conduits, 121
Construction phase, 3, 6
Construction safety, 11
Construction shafts, 264, 266
Construction surveys, 244-247
Continuity principle, 78-80
Contract conditions, 237-238
Contract drawings, 224-227

Contract forms, 237, 241
Contract preparation, 225-226
Contractor, 5
Contracts, 223-234, 237-239
Contracts, format, 226-234
Control points, 245-246
Controls, 108-109
Cooling water, 34-35
Corrosion protection, 59-66
Corrugated metal pipes, 220
Critical deflection limit, 214-215
Critical depth, 107-108
Critical flow, 76
Curved channels, 136-137
Curved sewers, 115-121

Darcy-Weisbach equation, 98-100
Demographic information needs, 16, 17, 23-24, 25-31
Design computation, 103-110
Design considerations, sulfides, 49-51, 56-57
Design criteria, infiltration, 35
Design criteria, inflow, 35
Design flow, 22-43
Design period, 21-22
Design phase, 3, 6
Distributed loads, 191-192
Draglines, 250-251
Drilling, excavation, 248-249
Drop manholes, 137-138
Ductile Iron Pipe (DIP), 157, 218-220

Elastomeric sealing compound pipe joints, 163
Embankment conditions, 175-182
Embankment conditions *(def.)*, 167
Embankments, 206-209
Embedment materials, 217-218
Encased pipe, 204-206
Energy concepts, 112
Energy losses, 82-86, 88-91
Energy principle, 80-82
Environmental factors, 177-178
Excavating equipment, tunnels, 264
Excavation, 248-251
Excessive excavation, 186
Exposed sewers, 125
External corrosion, 58-59

Federal funding, 7-8
Field data, contracts, 224-225
Field deflection, 215-216
Field strength, 206-209
Financial aspects, 16, 17-18
Financing, 7-10
Fixture-unit method of design, 40-43
Flap gates, 145

Flexible pipe, 119-121, 157-162, 197-198, 209-221
Float wells, 151-152
Flow depth, 74, 101-102, 122
Flow estimates, 103-107
Flow expansion, 74-75
Flow friction formulas, 93-100
Flow resistance, 93-102
Flow routing, 110
Flow situations, 70-78
Flow types, 70
Flow velocity, 122
Flow-slope relationships, 51-54
Foundations, 125, 194, 253-254
Friction coefficients, 100-102
Friction losses, 82-85
Front-end loaders, 251

Galvanic acton, 59
Gasket pipe joints, 163
Grade references, 245-246
Gravity earth forces, 166-174

Haunching, 195-196, 216-218, 221
Hazen-Williams formula, 97-98
Heat fusion pipe joints, 164
House connections, 139-141
Hydraulic computations, 103
Hydraulic design, 128-129
Hydraulic jump, 77-78, 108
Hydraulic jump *(def.)*, 76-77
Hydraulic principles, 70-73
Hydraulic symbols, 68-69

Induced trench sewer pipes, 179-182
Industrial areas, sewer needs, 33
Infiltration, 35, 36-38, 123
Infiltration *(def.)*, 21
Infiltration-exfiltration testing, 123
Inflow, 35-36, 71-73, 123
Inflow *(def.)*, 21
Information needs, 15-16
Information sources, 17-18
Installation, 202-204, 216-217, 221
Inspection for sulfides, 55-56
Institutional sewer needs, 34
Intercepting sewer *(def.)*, 2

Jacked sewer pipe, 183-186
Jacking, 259-260
Junctions, 137

Kutter's formula, 94-95

Laboratory load test, 213-214
Laboratory strength, 200-201
Land use, 24
Lateral sewer *(def.)*, 2
Line references, 245-246
Linings, 59-60
Load producing forces, 183-184, 187

Loading conditions, 167
Local funding, 8-10

Main sewer *(def.)*, 2
Manhole bench, 134-135
Manhole channel, 134-135
Manhole covers, 132-133
Manhole dimensions, 130-132
Manhole location, 114
Manhole steps, 133-134
Manhole to sewer connection, 133
Manhole walls, 132
Manning formula, 94-97, 100
Marston equation, 167-173, 177, 184-185, 215
Marston Theory, 166-167
Mastic pipe joints, 164
Measurement units, 12-14
Mechanical compression pipe joint, 163
Metering devices, 148-152
Minimum flow, 38-43
Mining methods, tunneling, 260-266
Momentum equation, 88-91
Momentum principle, 82

National Environmental Policy Act of 1969 (NEPA), 11-12
Negative projecting sewer pipe, 179-182, 209

Ocean outfalls, 146-147, 269-270
One-dimensional steady flow, 78
Open cut loads, 124
Open-channels, 71-72, 75-76, 80-81
Open-face mining, 261
Open-trench construction, 247-258
Outfall *(def.)*, 2
Outfall sewer *(def.)*, 2
Outfall sewers, 268-270
Outfalls, 146-148
Owner, 4-5

Partially filled pipe conditions, 54-55
Peak flow, 38-43
Performance limits, 196-200
Physical features, information needs, 15-16, 17
Pipe handling, 254
Pipe joints, 162-164
Pipe material, composition, 60-66
Pipe material, thickness, 60-66
Pipe placement, 254-255
Pipe quality, 254
Pipe stiffness, 213-214
Plastic sewer pipes, 212-217
Political aspects, 16, 17
Polyethylene (PE) pipe, 159-160
Polyvinyl Chloride (PVC) pipe, 160-161
Positive projecting sewer pipe, 175-179, 206-209
Precast concrete tunnel lining, 263-264

INDEX

Pressure-conduit flow, 72
Project development, 2-4
Project engineer, 5
Project investigation, 2-3, 6, 19-20
Project manuals, 234-237, 240-242
Project manuals, format, 235-236
Project manuals, purpose of, 234-235
Push-on pipe joint, 163

Railroad crossing, 266-267
Record keeping, construction, 270
Reinforced Plastic Mortar (RPM), 66
Reinforced Plastic Mortar (RPM) pipe, 162
Reinforced Thermosetting Resin (RTR), 66
Reinforced Thermosetting Resin (RTR) pipe, 161-162
Relief sewer *(def.)*, 1-2
Relief sewers, 128
Rigid pavement, loads, 192-193
Rigid pipes, 115-118, 154-156, 197, 200-209
Riverbank structures, 268
Rock trenches, 125

Safety, 10-11
Safety factors, 196-200
Safety, field procedures, 198-200
Sanitary sewer *(def.)*, 1
Sanitation needs, 16, 17
Sealing band joints, 164
Self-cleansing velocity, 105-107
Self-cleansing velocity *(def.)*, 105
Separate sewer *(def.)*, 2
Separate sewer system *(def.)*, 2
Separate sewers, 113
Service connections, 139-141
Service laterals, 139-141
Sewer depth, 121-122
Sewer plans, 229
Shallow manholes, 132
Siphons, 143-145
Site preparation, 247
Slope, 72-73
Slopes, 125, 182-183
Soil characteristics, 174-175, 179, 185-186
Soil classification, 217-218
Solvent cement pipe joints, 163-164
Specific energy, 86-88
Specifications, contracts, 238-239, 241-242
State funding, 8
Steady uniform flow, 71
Steel pipe, 157-158

Steel tunnel linings, 263
Steep slopes, 125
Storm sewer *(def.)*, 2
Strip map, 229
Stripping, excavation, 248
Subcritical flow *(def.)*, 76
Sulfide buildup forecasting, 51-55
Sulfide control, 57-66
Sulfide effects, 56-58
Sulfide generation, 47-51, 145
Supercritical flow *(def.)*, 76
Superimposed loads, 188-193
Surface restoration, 258
Surveys, 18-19, 244-247
System layout, 113-115

Terminal cleanouts, 138-139
Testing, 200-201, 123-124
Thermoplastic pipe, 158-161
Thermoset plastic pipe, 161-162
Timber tunnel linings, 261, 263
Traffic arteries, 267
Trench boxes, 199
Trench bracing, 251-253
Trench conditions, 167-175, 202-204, 209
Trench conditions *(def.)*, 167
Trench dewatering, 253
Trench dimensions, 247-248
Trench excavation, 248-251
Trench foundations, 253-254
Trench sheeting, 198-199, 251-253
Trenching, 249-251
Trenching machines, 249
Tributary areas, 24-25
Trunk sewer *(def.)*, 2
Tunnel construction, 246-247
Tunnel linig, 261-263
Tunnel shields, 260-261
Tunneling, 125, 258-266
Tunnels, 185-187

Use of santiary sewer systems, 7

Ventilation, 121
Ventilation, tunnel construction, 266
Vitrified clay pipe (VCP), 156

Wastewater flow measurement, 148-152
Waste water *(def.)*, 1
Wastewater *(def.)*, 1
Water-surface elevation, 107-109
Water-surface profiles, 91-93, 110
Waterway crossings, 267-268
Wave calculations, 110
Weirs, 148-151

Manning's

$$V = \frac{1.49}{n} R^{2/3} S^{1/2}$$

H-W:

$$\frac{h}{L_{ft}} = \frac{10.45}{D_{in}^{4.87}} \left(\frac{Q_{gpm}}{C}\right)^{1.85}$$

$$\frac{H-Z}{2.31} = P$$

Parallel $\quad h_1 = h_2$

$$Q_1 + Q_2 = Q_T$$

Soln 1: $\quad h_1 = K_1 |Q_1|^{0.85} Q_1$

Soln 2: $\quad K_1 (Q_T - Q_2)^{1.85} = K_2 Q_2^{1.85}$

$T_{fill} = V_0 / Q_{in}$

$T_{Empty} = V_0 / (Q_{out} - Q_{in})$

$h_{friction\ only} = 3.0 \times 10^{-5} Q^2$